Annals of Mathematics Studies

Number 158

# Green's Function Estimates for Lattice Schrödinger Operators and Applications

## J. Bourgain

PRINCETON UNIVERSITY PRESS

PRINCETON AND OXFORD

Copyright © 2005 by Princeton University Press

Published by Princeton University Press, 41 William Street, Princeton, New Jersey 08540

In the United Kingdom: Princeton University Press, 3 Market Place, Woodstock, Oxfordshire OX20 1SY

Library of Congress Control Number: 2004104492

ISBN: 691-12097-8 (hardcover); 0-691-12098-6 (paper)

British Library Cataloging-in-Publication Data is available

This book has been composed in LaTeX

The publisher would like to acknowledge the author of this volume for providing the camera-ready copy from which this book was printed.

Printed on acid-free paper.

pup.princeton.edu

Printed in the United States of America

10 9 8 7 6 5 4 3 2 1

# Contents

# Acknowledgment

The author is grateful to W. Schlag, Silvius Klein, and Ya-Ju Tsai for reading and comments on the initial versions of this manuscript.

The author would also like to acknowledge the National Science Foundation, which provided partial support for the research for this book with grant DMS 0322370.

# Green's Function Estimates for Lattice Schrödinger Operators and Applications

# Chapter One

## Introduction

We will consider infinite matrices indexed by $\mathbf{Z}$ (or $\mathbf{Z}^b$) associated to a dynamical system in the sense that

$$H = \left( H(x)_{m,n} \right)_{m,n \in \mathbf{Z}}$$

satisfies

$$H(x)_{m+1,n+1} = H(Tx)_{m,n}$$

where $x \in \Omega$, and $T$ is an ergodic measure-preserving transformation of $\Omega$. Typical settings considered here are

$$
\begin{array}{lll}
\Omega = \mathbf{T} & Tx = x + \omega & (1 - \text{frequency shift}) \\
\Omega = \mathbf{T}^d & Tx = x + \omega & (d - \text{frequency shift}) \\
\Omega = \mathbf{T}^2 & Tx = (x_1 + x_2, x_2 + \omega) & (\text{skewshift}) \\
\Omega = \mathbf{T}^2 & Tx = Ax, \text{ where } A \in SL_2(\mathbf{Z}), \text{hyperbolic}
\end{array}
$$

Thus

$$H(x)_{m,n} = \phi_{m-n}(T^m x) \qquad (1.0)$$

where the $\phi_k$ are functions on $\Omega$.

We will usually assume that $H(x)$ is self-adjoint, although many parts of our analysis are independent of this fact. Define

$$H_N = R_{[1,N]} H R_{[1,N]}$$

where $R_{[1,N]} = $ coordinate restriction to $[1, N] \subset \mathbf{Z}$, and the associated Green's functions are

$$G_N(E) = (H_N - E)^{-1}$$

(if $H_N - E$ is an invertible $N \times N$ matrix).

One of our concerns will be to obtain a 'good bound' on $G_N(E, x)$, except for $x$ in a "small" exceptional set. A typical statement would be the following:

$$\|G_N(E, x)\| < e^{N^{1-\delta}} \qquad (1.1)$$

and

$$|G_N(E, x)(m, n)| < e^{-c|m-n|} \text{ for } |m - n| > \frac{N}{10} \qquad (1.2)$$

for all $x$ outside a set of measure $< e^{-N^\sigma}$. Here $\delta, \sigma > 0$ are some constants. The exceptional set in $x$ does depend on $E$, of course. Such estimates are of importance in the following problems, for instance.

1. Spectral problems for lattice Schrödinger operators

Description of the spectrum Spec $H(x)$ and eigenstates of $H(x)$ (i.e., point spectrum, continuous (absolutely continuous or singular continuous) spectrum, localization, extended states, etc.)

2. Long-time behavior of linear time-dependent Schrödinger operators

$$i\frac{\partial u}{\partial t} + \Delta u + V(x,t)u = 0 \qquad (1.3)$$

The spatial variable $x \in \mathbf{T}^d$ (i.e., periodic bc).

The potential $V$ depends on time. It is well known that if $V$ is periodic in time (say, 1-periodic), we are led to study the monodromy operator

$$Wu(t) = u(t+1)$$

(which is unitary).

Again, the nature of spectrum and localization of eigenfunctions are key issues.

A well known example is the so-called kicked rotor problem

$$i\frac{\partial u}{\partial t} + a\frac{\partial^2 u}{\partial x^2} + ib\frac{\partial u}{\partial x} + \kappa\left[\cos x \sum_{n \in \mathbf{Z}} \delta(t-n)\right]u = 0 \qquad (1.4)$$

involving periodic "kicks" in time introduced as a model in quantum chaos. Here $V$ is discontinuous in time.

We assume $V$ real. We will also assume $V(\cdot, t)$ smooth in $x \in \mathbf{T}^d$ for all time. By the reality of $V$, there is conservation of the $L^2$-norm.

If $u_0 = u(0) \in H^s(\mathbf{T}^d)$, then

$$u(t) \in H^s \text{ for all time}$$

**Problem.** Possible growth of $\|u(t)\|_{H^s}$.

**Remark 1.** It turns out that in (1.4) with typical values of $a, b$ there is *almost-periodicity* in the following sense: Assume $u_0$ sufficiently smooth (depending on $s$). Then $u(t)$ is almost as periodic as an $H^s$-valued function and, in particular, $\sup_t \|u(t)\|_{H^s} < \infty$.

**Remark 2.** If in (1.3) we take $V$ also to be $t$-periodic, $u(t)$ is well known to be almost periodic in time as an $L^2$-valued function. But there are examples where $V$ is smooth in $x$ and $t$ and such that for some smooth initial data $u_0$

$$\sup_t \|u(t)\|_{H^s} = \infty \text{ for all } s > 0$$

3. KAM-theory via the Nash-Moser method

We refer here to a method developed by W. Craig, G. Wayne, and myself to construct quasi-periodic solutions of nonlinear Hamiltonian PDEs. This approach was used originally as a substitute of the usual KAM-scheme (as used in this context by S. Kuksin) in situations involving multiplicities or near-multiplicities of normal frequencies. These always appear, except in 1D problems with Dirichlet boundary conditions. It was realized later that this technique is also of interest in the "classical context" involving finite-dimensional phase space (leading, for instance, to a Melnikov-type result with the "right" nonresonance assumptions) and applies in certain non-Hamiltonian settings.

If we follow a Newton-type iteration scheme, the basic difficulty is the inversion of nondiagonal operators obtained by linearizing the (nonlinear) PDE.

Consider, for instance, the Schrödinger case

$$iu_t + \Delta u + \varepsilon F(u, \bar{u}) = 0 \tag{1.5}$$

The linearized operator expressed in Fourier modes then becomes

$$T = D + \varepsilon S$$

where $D$ is diagonal with diagonal elements of the form

$$D_{k,n} = k.\omega + \mu_n = k.\omega + |n|^2 + 0(1) \quad (k \in \mathbf{Z}^b, n \in \mathbf{Z}^d) \tag{1.6}$$

and $S$ is a Toeplitz-type matrix with (very) smooth symbol, i.e.,

$$S\big((k, n), (k', n')\big) = \hat{\varphi}(k - k', n - n')$$

where $\hat{\varphi}(\xi)$ decays rapidly for $|\xi| \to \infty$.

In (1.6), $b$ = dimension of invariant tori, and $\omega \in \mathbf{R}^b$ is the frequency vector. The matrix $T$ is finite (depending on the iteration step), and we seek appropriate bounds on $T^{-1}$. The problem again involves small-divisor issues and is treated by multiscale analysis.

Returning to $H(x)$, one important special case is given by

$$H(x) = \lambda.v(T^n x)\delta_{nn'} + \Delta \tag{1.7}$$

where $\Delta$ is the usual lattice Laplacian

$$\Delta(n, n') = 1 \text{ if } |n - n'| = 1$$
$$= 0 \text{ otherwise}$$

Letting $v(x) = \cos x$ on $\mathbf{T}$, $Tx = x + \omega$ = shift, we obtain the Almost Mathieu operator

$$H_\lambda(x) = \lambda \cos(x + n\omega) + \Delta \tag{1.8}$$

introduced by Peierls and Hofstadter in the study of a Bloch electron in a magnetic field and studied extensively afterwards by many authors.

For (1.8), there is basically a complete understanding of the nature of the spectrum. Assume that $\omega$ satisfies a diophantine condition

$$\text{dist}\,(k\omega, 2\pi\mathbf{Z}) = \|k.\omega\| > c|k|^{-C} \text{ for } k \in \mathbf{Z}\backslash\{0\}$$

Then, for a.e. $x$,

(i) $\underline{\lambda > 2}$: $H(x)$ has p.p. spectrum

(ii) $\underline{\lambda = 2}$: $H(x)$ has purely s.c. spectrum

(iii) $\underline{\lambda < 2}$: $H(x)$ has purely a.c. spectrum

Thus there is a phase transition at $\lambda = 2$.

This model has a special and remarkable self-duality property (wrt Fourier transform)

$$\cos \rightarrow \tfrac{1}{2}\Delta$$
$$\Delta \rightarrow 2\cos$$

observed and exploited first by Aubry. One of its implications is that

$$\operatorname{Spec} H_\lambda = \operatorname{Spec} H_{\frac{4}{\lambda}}$$

(referring to the "topological spectrum" that is independent of $x$).

In more general situations involving shifts,

$$\lambda v(x + n\omega)\delta_{nn'} + \Delta \tag{1.9}$$

with $v$ real analytic on $\mathbf{T}^d$, a rough picture is the following:

$\lambda$ **large**: p.p. spectrum with Anderson localization

$\lambda$ **small**: purely a.c. spectrum

$\lambda$ **intermediate**: possible coexistence of different spectral types

Recall that Anderson localization means the following:

Assume $\psi$ an extended state, i.e.,

$$H\psi = E\psi \ \text{ and } \ |\psi_n| \lesssim |n|^C$$

Then $\psi \in \ell^2$ and

$$|\psi_n| < e^{-c|n|} \ \text{ for } |n| \rightarrow \infty$$

(in particular, $E$ is an eigenvalue).

Related to possible coexistence of different spectral types (in various energy regions), one may prove the following:

Consider

$$H = (\lambda \cos n\omega_1 + \tau \cos n\omega_2)\delta_{nn'} + \Delta \tag{1.10}$$

where $\lambda < 2$, and $\tau$ is small. Then, for $\omega = (\omega_1, \omega_1)$ in a set of positive measure, $H$ has both point spectrum and a.c. spectrum.

**Remark.** If in (1.9) we replace the shift by the skew shift, one expects a different spectral behavior with localization for all $\lambda > 0$ (as is the case of a random potential).

This problem is open at this time. It is known that for all $\lambda > 0$ and $\omega$ in a set of positive measure

$$H = \lambda \left( \cos \frac{n(n-1)}{2}\omega \right)\delta_{nn'} + \Delta$$

has some p.p. spectrum.

This text originates from lectures given at the University of California, Irvine, in 2000 and UCLA in 2001. The first 17 chapters deal mainly with localization problems for quasi-periodic lattice Schrödinger operators. Part of this material is borrowed from the original research papers. However, we did revise the proofs in order to present them in a concise form with emphasis on the key analytical points. The main interest, independent of style, is that we give an overview of a large body of research presently scattered in the literature. The results in Chapter 8 on regularity properties of the Lyapounov exponent and Integrated Density of states (IDS) are new. They refine the work from [G-S] described in Chapter 7. (Nonperturbative quasi-periodic localization is discussed in Chapter 10. We follow the paper [B-G] but also treat the general multifrequency case (in 1D). In [B-G], only the case of two frequencies was considered. Our presentation here uses the full theory of semialgebraic sets and in particular the Yomdin-Gromov uniformization theorem. This material is discussed in Chapter 9.

Chapters 18, 19, and 20 deal with the problem of constructing quasi-periodic solutions for infinite-dimensional Hamiltonian systems given by nonlinear Schrödinger (NLS) or nonlinear wave equations (NLW). Earlier research, mainly due to C. Wayne, S. Kuksin, W. Craig, and myself (see [C] for a review), left open a number of problems. Roughly, only 1D models and the 2D NLS could be treated.

In this work we develop a method to deal with this problem in general. Thus we consider NLS and NLW (with periodic boundary conditions) given by a smooth Hamiltonian perturbation of a linear equation with parameters and proof persistency of a large family of smooth quasi-periodic solutions of the linear equation. This is achieved in arbitrary dimension. Compared with earlier works, such as [C-W] and [B1], we do rely here on more powerful methods to control Green's functions. These methods were developed initially to study quasi-periodic localization problems. Thus the material in Chapters 18 to 20 is also new.

We want to emphasize that it is our only purpose here to convey a number of recent developments in the general area of quasi-periodic localization and the many remaining problems. This is an ongoing area of research, and our understanding of most issues is still far from fully satisfactory. The material discussed, moreover, covers only a portion of these developments (for instance, we don't discuss at all renormalization methods, as initiated by B. Hellfer and J. Sjostrand). We have largely ignored the historical perspective. Nevertheless, it should be pointed out that this field to a large extent owes its existence to the seminal work of Y. Sinai and his collaborators (in particular, the papers [Si], [C-S], and [D-S]), as well as the paper [F-S-W] by Frohlich, Spencer, and Wittwer. One of the significant differences, however, between these works (and some later developments such as [E]) and ours on the technological side is the fact that we don't rely on eigenvalue parametrization methods, which seem, in particular, very hard to pursue in multidimensional problems (such as considered in [B-G-S], for instance). It turns out that, as mentioned earlier, lots of the analysis is independent of self-adjointness and has potential applications to non-self-adjoint problems. We rely heavily in both perturbative and nonperturbative settings on methods from subharmonic function theory and the theory of semianalytic sets, which somehow turn out to be more "robust" than eigenvalue techniques (the results obtained are a bit weaker in the sense that "good" frequencies are not always characterized by diophantine conditions, as in [Si], [F-S-W], [E], or [J]). Jitomirskaya's paper [J] certainly underlies much of this recent research. Besides settling the spectral picture for the Almost Mathieu operator and the phase transition mentioned earlier, it initiated the nonperturbative approach with emphasis on the Lyapounov exponent and transfer matrix. Some parts of the analysis were restricted to the cosine potential, and the extension to general polynomial or real analytic potentials (see [B-G]) lies at the root of the material presented in these notes.

Next, a bit more detailed discussion of the content of the different chapters. Chapters 2 through 11 are closely related to the papers [B-G] and [G-S] on nonperturbative localization for quasi-periodic lattice Schrödinger operators of the form

$$H_x = \lambda v(x + n\omega) + \Delta \tag{1.11}$$

where $v$ is a real analytic potential on $\mathbf{T}^d$ ($d = 1$ or $d > 1$), and $\Delta$ denotes the lattice Laplacian on $\mathbf{Z}$. We are mainly concerned with the issues of pure point spectrum, Anderson localization, dynamical localization, and regularity properties of the IDS. A key ingredient is the positivity of the Lyapounov exponent for sufficiently large $\lambda$. The results are nonperturbative in the sense that the condition $\lambda > \lambda_0(v)$ depends on $v$ only and not on the arithmetical properties of the rotation vector $\omega$ (provided we assume $\omega$ to satisfy some diophantine condition).

Here and throughout this exposition, extensive use is made of subharmonic function techniques and the theory of semialgebraic sets. A summary of certain basic results in semialgebraic set theory appears in Chapter 9. The basic localization theorem is proven in Chapter 10, and some extensions of the method to more general operators are given in Chapter 11.

In Chapter 12 we recall some elements from Kotani's theory for later use. But this is far from a complete treatment of this topic, and several other results and aspects are not mentioned.

In Chapter 13 we exhibit point spectrum in certain two-frequency models of the form (0.11) with small $\lambda$. This fact shows that, contrary to the localization theory, the nonperturbative results on absolutely continuous spectrum, as obtained in [B-J] for one-frequency models, fail in the multifrequency case. Equivalently, invoking the Aubry duality, the quasi-periodic localization results on the $\mathbf{Z}^2$-lattice (as discussed in Chapter 17) are only perturbative.

In Chapter 14 we develop a general perturbative method to control Green's functions of certain lattice Schrödinger operators. The main result is in some way an "analogue" of Cartan's theorem in analytic function theory for holomorphic matrix-valued functions.

This approach has a wide range of applications. First, it allows us to control Green's functions for general Jacobi operators of the form (1.0) associated to a dynamical system given by a skew shift (Chapter 15). As an application, we prove the almost periodicity of smooth solutions of the kicked rotor equation (1.4) with small $\kappa$ and typical parameter values $a, b$ (Chapter 16). Next, an extension of Chapter 14 to a 2D setting permits us to establish Anderson localization for operators of the form (1.11) on the $\mathbf{Z}^2$-lattice. The statement is perturbative, i.e., $\lambda > \lambda_0(v, \omega)$. However, as indicated earlier, a nonperturbative result may be false in this situation. In fact, considering the multifrequency generalizations of the Almost-Mathieu operator

$$H_x = \lambda\big(\cos(x_1 + n\omega_1) + \cos(x_2 + n\omega_2)\big) + \Delta \quad \text{(on } \mathbf{Z}) \tag{1.12}$$

and its "dual"

$$\widetilde{H}_\theta = \cos(\theta + n_1\omega_1 + n_2\omega_2) + \frac{\lambda}{4}\Delta \quad \text{(on } \mathbf{Z}^2) \tag{1.13}$$

it turns out that for arbitrary $\lambda > 0$, there is a set of frequencies $\Omega = \Omega_\lambda \subset \mathbf{T}^2$ of small but positive measure such that for $\omega \in \Omega$ and $x$ in a set of positive measure, we have

$$\text{mes}\left(\sum_{pp} H_x\right) > 0$$

(in fact, there may be coexistence of different spectral types here). Hence $\widetilde{H}_\theta$ has true (i.e., not $\ell^2$) extended states for almost all $\theta$. ($\mathbf{Z}^\ell$-operators of the form (1.13) were first studied in [C-D].)

Finally, the method from Chapter 14 enable us to treat KAM-type problems via the Lyapounov-Schmidt approach (see [C-W]) in a number of situations that, due to large sets of resonances, seemed untractable previously. (Typical issues left open here from the earlier works are the NLS in space dimension $D \geq 3$ and the NLW in space dimension $D \geq 2$).

In Chapter 18 we give a new proof of Melnikov's theorem on persistency of $b$-dimensional tori in (finite-dimensional) phase space of dimension $> 2b$ (for Hamiltonian perturbations of a linear system, assuming the Hamiltonian given by a polynomial.) The spirit of the argument is closely related to earlier discussion on perturbative localization. In particular, semialgebraic set theory is used again to restrict the parameter space.

In Chapters 19 and 20 we then apply this scheme to obtain quasi-periodic solutions for nonlinear PDE (with periodic bc), thus involving an infinite-dimensional phase space. Chapter 19 deals with NLS and Chapter 20 with NLW. Compared with the finite-dimensional phase space setting discussed in Chapter 18, there are some additional difficulties (due to large sets of resonant normal modes). But the method is sufficiently robust to deal with them. An additional ingredient involved here is a "separated cluster structure" for the near-resonant sets (noticed first by T. Spencer in a 2D-Schrödinger context).

As mentioned earlier, results from Chapters 18 to 20 treat only perturbations of linear systems with parameters. Starting from a genuine nonlinear problem, this format may often be reached through the theory of normal forms and amplitude-frequency modulation (see [K-P] and [B2]). This is a different aspect of the general problem, however, that is not addressed here.

In the Appendix we consider lattice Schrödinger operators associated to strongly mixing dynamical systems. We mainly summarize results from [C-S] and [B-S] based on the Figotin-Pastur approach. So far, this method to evaluate Lyapounov exponent has succeeded only in a strongly mixing context.

# References

[B1] J. Bourgain. Quasi-periodic solutions of Hamiltonian perturbations of $2D$-linear Schrödinger equations, *Annals of Math.* 148 (1998), 363–349.

[B2] J. Bourgain. Construction of periodic solutions of nonlinear wave equations in higher dimension, *GAFA* 5 (1995), 629–639.

[G-B] J. Bourgain, M. Goldstein. On non-perturbative localization with quasi-periodic potential, *Annals of Math.* (2) 152(3) (2000), 835–879.

[BGS] J. Bourgain, M. Goldstein, W. Schlag. Anderson localization for Schrödinger operators on $\mathbf{Z}^2$ with quasi-periodic potential, *Acta Math.* 188 (2002), 41–86.

[B-J] J. Bourgain, S. Jitomirskaya. Nonperturbative absolutely continuous spectrum for 1D quasi-periodic operators, preprint 2000.

[B-S] J. Bourgain, W. Schlag. Anderson localization for Schrödinger operators on $\mathbf{Z}$ with strongly mixing potentials, *CMP* 215 (2000), 143–175.

[C-D] V. Chulaevsky, E. Dinaburg. Methods of KAM theory for long-range quasi-periodic operators on $\mathbf{Z}^n$. Pure point spectrum, *CMP* 153(3) (1993), 539–557.

[C-S] V. Chulaevsky, T. Spencer. Positive Lyapounov exponents for a class of deterministic potentials, *CMP* 168 (1995), 455–466.

[C-Si] V. Chulaevsky, Y. Sinai. Anderson localization for multifrequency quasi-periodic potentials in one dimension, *CMP* 125 (1989), 91–121.

[C] W. Craig. Problemes de petits diviseurs dans les Equations aux dérivés partielles, *Panoramas et Synthèses*, 9, SMF, Paris, 2000.

[C-W] W. Craig, C. Wayne. Newton's method and periodic solutions of nonlinear wave equations, *CPAM* 46 (1993), 1409–1501.

[D-S] E.I. Dinaburg, Y. Sinai. Methods of KAM-theory for long-range quasi-periodic operators on $\mathbf{Z}^\nu$. Pure point spectrum, *CMP* 153(3) (1993), 559–577.

[E] L.H. Eliasson. Discrete one-dimensional quasi-periodic Schrödinger operators with pure point spectrum, *Acta Math.* 179 (1997), 153–196.

[F-S-W] J. Fröhlich, T. Spencer, P. Wittwer. Localization for a class of one dimensional quasi-periodic Schrödinger operators, *Comm. Math. Phys.* 132 (1990), 5–25.

[G-S]  M. Goldstein, W. Schlag. Hölder continuity of the integrated density of states for quasi-periodic Schrödinger equations and averages of shifts of sub-harmonic functions, *Annals of Math.* 154 (2001) 155–203.

[J]  S. Jitomirskaya. Metal-insulator transition for the Almost Mathieu operator, *Annals of Math.* (2) 150(3) (1999), 1159–1175.

[K-P]  S. Kuksin, J. Pöschel. Invariant Cantor manifolds of quasi-periodic oscillations for a nonlinear Schrödinger equation, *Annals of Math.* 143 (1996), 149–179.

[Si]  Ya. G. Sinai. Anderson localization for one-dimensional difference Schrödinger operator with quasi-periodic potential, *J. Statist. Phys.* 46 (1987), 861–909.

# Chapter Two

## Transfer Matrix and Lyapounov Exponent

Consider 1D lattice Schrödinger operators of the form

$$H = v(T^j x)\delta_{jj'} + \Delta$$

Assume that $\psi = (\psi_j)_{j \in \mathbf{Z}}$ is a sequence satisfying

$$H\psi = E\psi$$

Then

$$\begin{pmatrix} \psi_{n+1} \\ \psi_n \end{pmatrix} = M_n(E) \begin{pmatrix} \psi_1 \\ \psi_0 \end{pmatrix}$$

where

$$M_n(E) = M_n(E, x) = \prod_{j=n}^{1} \begin{pmatrix} v(T^j x) - E & -1 \\ 1 & 0 \end{pmatrix} \tag{2.1}$$

is the transfer (or fundamental) matrix.
Define further

$$L_N(E) = \frac{1}{N} \int \log \|M_N(x, E)\| dx \tag{2.2}$$

and

$$\begin{aligned} L(E) &= \lim_{N \to \infty} L_N(E) \\ &= \text{Lyapounov exponent} \end{aligned} \tag{2.3}$$

Observe that by submultiplicativity

$$\|M_{n_1+n_2}(x, E)\| \leq \|M_{n_2}(T^{n_1} x, E)\| . \|M_{n_1}(x, E)\|$$

hence

$$\begin{aligned} L_{n_1+n_2}(E) &\leq \tfrac{n_1}{n_1+n_2} L_{n_1}(E) + \tfrac{n_2}{n_1+n_2} L_{n_2}(E) \\ L(E) &= \lim_{n \to \infty} L_n(E) \text{ exists} \end{aligned}$$

and by Kingman's ergodic theorem (assuming $T$ ergodic) (see [K])

$$L(E) = \lim_{n \to \infty} \frac{1}{n} \log \|M_n(x, E)\| \qquad x \text{ a.e.}$$

There is the following relation between $M_n(E)$ and determinants

$$M_n(x, E) = \begin{bmatrix} \det(H_n(x) - E) & -\det(H_{n-1}(Tx) - E) \\ \det(H_{n-1}(x) - E) & -\det(H_{n-2}(Tx) - E) \end{bmatrix} \tag{2.4}$$

Define the *integrated density of states* as

$$N(E) = \lim_{n \to \infty} \frac{1}{n} \, \# \left( ] - \infty, E] \cap \operatorname{Spec} H_n(x) \right) \qquad x \text{ a.e.}$$

The relation to the Lyapounov exponent is expressed by the Thouless formula (see [S], for instance)

$$L(E) = \int \log |E - E'| dN(E') \tag{2.5}$$

The convergence $\frac{1}{N} \log \|M_N(x, E)\| \to L(E)$ can be made more precise in certain cases by exploiting the specific structure (in particular, the transformation $T$). In what follows, an important role will be played by large deviation theorems (LDT) of the form

$$\operatorname{mes}[x \in \mathbf{T}^d \left| \, \left| \frac{1}{N} \log \|M_N(x, E)\| - L_N(E) \right| > \kappa \right] < \delta(N, \kappa) \tag{2.6}$$

where $\delta(N, \kappa) \overset{N \to \infty}{\to} 0$ (fixing $\kappa > 0$).

This bound $\delta(N, \kappa)$ will usually be exponential in $N$; thus

$$\delta(N, \kappa) < e^{-N^\sigma} (\sigma > 0)$$

for $N$ large enough. Such estimates are of particular interest in estimating Green's functions by Cramer's rule.

One has for $1 \le n_1 \le n_2 \le N$, by (2.4)

$$
\begin{aligned}
|G_N(E, x)(n_1, n_2)| &= |(H_N(x) - E)^{-1}|(n_1, n_2)| \\
&= \frac{|\det[H_{n_1-1}(x) - E]| \, |\det[H_{N-n_2}(T^{n_2}x) - E]|}{|\det[H_N(x) - E]|} \\
&\le \frac{\|M_{n_1}(x, E)\| \, \|M_{N-n_2}(T^{n_2}x, E)\|}{|\det[H_N(x) - E]|}
\end{aligned}
\tag{2.7}
$$

Assume that

$$L_{N_0}(E) < L(E) + \kappa$$

and

$$\left| \frac{1}{n} \log \|M_n(y, E)\| - L_n(E) \right| < \kappa$$

for $N_0 \le n \le N$ and $y \in \{x, Tx, \dots, T^N x\}$.

Then

$$(2.7) < \frac{e^{(N - |n_1 - n_2|) L(E) + 2\kappa N + 0(N_0)}}{|\det[H_N(x) - E]|} \tag{2.8}$$

Returning to (2.4), if we allow replacement of $N$ by $N - 1$ or $N - 2$ and $x$ by $Tx$, we may replace the denominator in (2.8) by $\|M_N(x, E)\| > e^{NL(E) - 2\kappa N}$. Thus we obtain

$$|G_\Lambda(E, x)(n_1, n_2)| < e^{-L(E)|n_1 - n_2| + 0(\kappa N + N_0)} \tag{2.9}$$

where $\Lambda$ is one of the intervals

$$[1, N], \ [1, N - 1], \ [2, N], \ [2, N - 1]$$

It is clear from (2.9) that positivity of the Lyapounov exponent

$$L(E) > c > 0$$

is important to get decay estimates on the Green's function. A major advantage of this technique is that it may provide nonperturbative results.

# References

[K]   J.F.C. Kingman.  The ergodic theory of subadditive stochastic processes, *J. Royal Statist. Soc.* B30 (1968).

[S] J. Avron, B. Simon. Almost periodic Schrödinger operators II. The integrated density of states, *Duke Math. J.* 50 (1) (1983), 369–391.

# Chapter Three

## Herman's Subharmonicity Method

There is a particularly simple method to obtain lower bounds on $L(E)$ in case $v(x)$ is a trigonometric polynomial. The argument is based on Jensen's inequality. We consider the example $v(x) = \cos x$.

**Proposition 3.1.** *Consider*

$$H(x) = \lambda \cos(n\omega + x)\delta_{nn'} + \Delta$$

*Then*

$$L(E) \geq \log \frac{\lambda}{2} \tag{3.2}$$

**Proof.** Write

$$
\begin{aligned}
L_N(E) &= \tfrac{1}{N} \int \log \left\| \prod_N^1 \begin{pmatrix} \lambda \cos(\theta + j\omega) - E & -1 \\ 1 & 0 \end{pmatrix} \right\| d\theta \\
&= \tfrac{1}{N} \int_{|z|=1} \log \left\| \prod_N^1 \begin{pmatrix} \frac{\lambda}{2} e^{-ij\omega} + \frac{\lambda}{2} e^{ij\omega} z^2 - Ez & -z \\ z & 0 \end{pmatrix} \right\| \\
&\geq \tfrac{1}{N} \log \left\| \begin{pmatrix} \frac{\lambda}{2} e^{-ij\omega} & 0 \\ 0 & 0 \end{pmatrix} \right\| \quad \text{(by Jensen)} \\
&= \log \tfrac{\lambda}{2}.
\end{aligned}
$$

**Remarks.**
**1.** The argument clearly generalizes to trigonometric polynomials.
**2.** For $v(x) = \cos x$, $L(E) \geq \log \frac{\lambda}{2}$ is optimal as energy-independent lower bound

It follows in particular that $L(E) > c > 0$ for all $E$ if $\lambda > 2$ (which is the regime of p.p. spectrum and Anderson localization). If $v$ is given by a real analytic function on $\mathbf{T}$, i.e.,

$$v(\theta) = \sum_{k \in \mathbf{Z}} \hat{v}(k) e^{2\pi i k\theta} \quad |\hat{v}(k)| < e^{-\rho|k|}$$

then it is still possible to obtain a lower bound by subharmonicity.

**Proposition 3.3.** *Let* $H(x) = \lambda v(n\omega + x)\delta_{nn'} + \Delta$ *with* $v$ *as above, $v$ nonconstant. Then, for* $\lambda > \lambda_0$,

$$L(E) > \frac{1}{2} \log \lambda$$

*where* $\lambda_0 = \lambda_0(v)$.

This result is due to Sorets-Spencer [S-S]. Observe that in this issue, approximation of $v$ by trigonometric polynomials and use of Herman's argument fails unless additional assumptions on $v$ are made.

Complexify $v$ to the strip $|\mathrm{Im}\, z| < \frac{\rho}{10}$, i.e.,

$$v(z) = \sum_{k \in \mathbf{Z}} \hat{v}(k) e^{2\pi i k z}$$

satisfying

$$|v(z)| \le \sum |\hat{v}(k)| e^{2\pi |k|\, |\mathrm{Im}\, z|}$$

$$\le \sum e^{-\rho|k| + \frac{2\pi}{10}\rho|k|} < C$$

Consider the holomorphic matrix-valued function on $|\mathrm{Im}\, z| < \frac{\rho}{10}$

$$M_n(z, E) = \prod_n^1 \begin{pmatrix} \lambda v(z + j\omega) - E & -1 \\ 1 & 0 \end{pmatrix}$$

for which

$$\| M_n(z, E) \| < (C|\lambda| + |E| + 1)^n$$

Thus

$$u(z) = \frac{1}{n} \log \| M_n(z, E) \| \tag{3.4}$$

is subharmonic and satisfies

$$0 \le u(z) \le \log(1 + |E| + C|\lambda|) \tag{3.5}$$

We also will need the following fact:

For all $\delta > 0$, there is $\varepsilon > 0$ such that

$$\inf_{E_1} \sup_{\frac{\delta}{2} < y < \delta} \inf_{x \in [0,1]} |v(x + iy) - E_1| > \varepsilon \tag{3.6}$$

This may be seen immediately by the compactness argument (otherwise, for some $E_1, v(z) - E_1$ would have infinitely many zeros; hence $v \equiv E_1$).

Fix $0 < \delta \ll \rho$, and let $\varepsilon$ be as in (3.6).

There is some $\frac{\delta}{2} < y_0 < \delta$ such that

$$\inf_{x \in [0,1]} \left| v(x + iy_0) - \frac{E}{\lambda} \right| > \varepsilon$$

and, by periodicity of $v_1$, also

$$\inf_{x \in \mathbf{R}} \left| v(x + iy_0) - \frac{E}{\lambda} \right| > \varepsilon \tag{3.7}$$

Define

$$\lambda_0 = 10\varepsilon^{-10}$$

and take $\lambda > \lambda_0$.

It follows from (3.7) that

$$\| M_n(iy_0, E) \| \ge |\langle M_n(iy_0, E) \begin{pmatrix} 1 \\ 0 \end{pmatrix}, \begin{pmatrix} 1 \\ 0 \end{pmatrix} \rangle|$$

$$> (\lambda\varepsilon - 1)^n$$

Hence

$$u(iy_0) > \log(\lambda\varepsilon - 1) \tag{3.8}$$

Denote $\mu \in \mathcal{M}([y = 0] \cup [y = \frac{\rho}{10}])$, the harmonic measure of $y_0$ in the strip $0 \le y \le \frac{\rho}{10}$

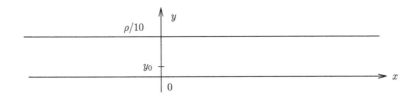

Clearly,

$$\mu[y = \tfrac{\rho}{10}] \;<\; \tfrac{10y_0}{\rho} \;<\; \tfrac{10\delta}{\rho}$$

$$\left.\frac{d\mu}{dx}\right|_{y=0} \;<\; \tfrac{y_0}{x^2+y_0^2}$$

By subharmonicity, (3.5) and (3.6),

$$\log(\lambda\varepsilon - 1) \;<\; u(iy_0) \le \int_{y=0} u(x)\mu(dx) + \int_{y=\frac{\rho}{10}} u(x+iy)\mu(dx)$$

$$< \int_{\mathbf{R}} u(x)\tfrac{y_0}{x^2+y_0^2}dx + \tfrac{10\delta}{\rho}\left[\begin{array}{c}\sup \\ |\operatorname{Im} z| = \tfrac{\rho}{10}\end{array} u(z)\right]$$

$$< \tfrac{2}{\delta}\int_0^1 u(x)dx + \tfrac{10\delta}{\rho}(\log C\lambda)$$

(since $u$ is 1-periodic in $x$), and therefore,

$$\frac{2}{\delta}\int_0^1 u(x)dx > \frac{1}{2}\log\lambda$$

A better estimate is obtained by writing

$$\int_{\mathbf{R}} u(x+a)\frac{y_0}{x^2+y_0^2}dx > \frac{1}{2}\log\lambda$$

with $a \in \mathbf{R}$ arbitrary; see (3.7).

Integrating in $a \in [0,1]$ thus implies

$$\int_0^1 u(x)dx > \frac{1}{2}\log\lambda \quad (\lambda > \lambda_0)$$

This proves Proposition 3.3.

**Remarks.**
**1.** Both Herman's [H] and Soretz-Spencer's [S-S] arguments are independent of diophantine assumptions on $\omega$.
**2.** For $v = v(x)$ real analytic on $\mathbf{T}^d, d > 1$, and considering the multifrequency shift by $\omega$, Proposition 3.3 holds if we make a diophantine assumption on $\omega$ (see [B-G] and [G-S] in Chapter 1).

The exact analogue of Proposition 3.3 for $d > 1$ as a uniform minoration with no dependence on $\omega$ was proven more recently in [Bo].
**3.** A natural question is whether a Soretz-Spencer type theorem holds in Gevrey class.

The key point in the proof of Proposition 3.3 is to avoid the set $[|v| \approx 0]$ by complexification. This procedure does not work for $d > 1$ unless again additional assumptions on $v$ are made. The argument in [Bo] uses diophantine considerations, although the final conclusion is independent of them.

# References

[Bo]  J. Bourgain. Positivity and continuity of the Lyapounov exponent for shifts on $\mathbf{T}^d$ with arbitrary frequency vector and real analytic potential, preprint 2002.

[H]  M. Herman.  Une methode pour minorer les exposants de Lyapounov et quelques examples montrant le caractère local d'un théorème d'Arnold et de Moser sur le tore de dimension 2, *Comment. Math. Helv.* 58 (1983), 453–562.

[S-S]  E. Soretz, T. Spencer. Positive Lyapounov exponents for Schrödinger operators with quasi-periodic potentials, *CMP* 142(3) (1991), 543–566.

# Chapter Four

## Estimates on Subharmonic Functions

The material presented in this chapter appears in [B-G], [G-S] (see Chapter 1), and [B-G-S] with slightly different formulations and proofs.

Assume $u = u(x)$ 1-periodic with subharmonic extension $\tilde{u} = \tilde{u}(z)$ to the strip $|\text{Im } z| < 1$ satisfying

$$|u| \le 1, \quad |\tilde{u}| \le B \tag{4.1}$$

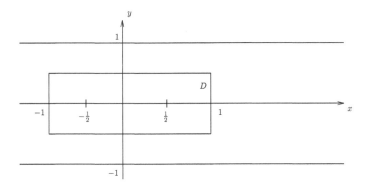

Apply Riesz representation on $D$ as above.

Thus in particular for $|x| \le \frac{3}{4}$,

$$u(x) = \int \log |x - w| \mu(dw) + h(x) \tag{4.2}$$

where

$$\frac{d\mu}{dw} = \Delta \tilde{u}$$

Hence $\mu$ is a positive measure on $D$, and $h$ is harmonic on $D$.

By (4.1),

$$\|\mu\| \lesssim B \tag{4.3}$$

$$|\partial_x^{(\alpha)} h| \lesssim B \text{ for } |x| < \frac{3}{4} \tag{4.4}$$

Denote

$$v(x) = \int_D \log |x - w| \mu(dw)$$

Hence

$$\partial_x v = \int \frac{x - \operatorname{Re} w}{(x - \operatorname{Re} w)^2 + (\operatorname{Im} w)^2} \mu(dw)$$

$$= \mathcal{H}[\nu] \tag{4.5}$$

($\mathcal{H}$ = Hilbert-transform) where $\nu \geq 0$ is the measure on $\mathbf{R}$ given by

$$\frac{d\nu}{dx} = \int \frac{|\operatorname{Im} w|}{(x - \operatorname{Re} w)^2 + (\operatorname{Im} w)^2} \mu(dw)$$

Hence

$$\|\nu\| \leq \|\mu\| \lesssim B \tag{4.6}$$

**Corollary 4.7.** $|\hat{u}(k)| \lesssim \frac{B}{|k|}$.

**Proof.** Take smooth $\eta$, supp $\eta \subset \left[ -\frac{3}{4}, \frac{3}{4} \right]$, $\sum_{j \in \mathbf{Z}} \eta(x + j) = 1$. Then

$$\hat{u}(k) = \int_{\mathbf{R}} u(x) \eta(x) e^{-2\pi i k x} dx$$

$$|\hat{u}(k)| \sim \frac{1}{|k|} |\widehat{\partial_x(u\eta)}(k)|$$

$$\leq \frac{1}{|k|} \left( |\widehat{\partial_x(v\eta)}(k)| + |\widehat{\partial_x(h\eta)}(k)| \right)$$

$$\lesssim \frac{B}{|k|}$$

by (4.4), (4.5), and (4.6).

**Corollary 4.8.** *Assume, moreover, that*

$$\operatorname{mes}[x \in \mathbf{T} \mid |u(x) - \langle u \rangle| > \varepsilon_0] < \varepsilon_1$$

*Then*

$$\|u\|_{BMO(\mathbf{T})} \leq C(\varepsilon_0 + \sqrt{\varepsilon_1 B}) \tag{4.9}$$

**Proof.** We may assume $\langle u \rangle = 0$. From the hypothesis

$$u = u_0 + u_1$$

$$\|u_0\|_\infty \leq \varepsilon_0, \|u_1\|_1 < \varepsilon_1$$

Let

$$\varepsilon \sim \sqrt{\frac{\varepsilon_1}{B}}$$

and denote $P_\varepsilon(t) = \frac{1}{\varepsilon} P(\frac{t}{\varepsilon})$, with $P \geq 0$ compactly supported, $\int P = 1$.

Since $u = v + h = u_0 + u_1$, write

$$u = (u - P_\varepsilon u) + P_\varepsilon u_0 + P_\varepsilon u_1$$

$$= (v - P_\varepsilon v) + (h - P_\varepsilon h) + P_\varepsilon u_0 + P_\varepsilon u_1$$

where

$$\|h - P_\varepsilon h\|_\infty \lesssim \varepsilon B \quad \text{by (4.4)}$$

$$\|P_\varepsilon u_0\|_\infty < \varepsilon_0$$

$$\|P_\varepsilon u_1\|_\infty \lesssim \frac{\varepsilon_1}{\varepsilon}$$

and

$$\partial_x v = \mathcal{H}[\nu]$$
$$\partial_x (v - P_\varepsilon v) = \mathcal{H}[\nu - P_\varepsilon \nu]$$

Thus

$$\|v - P_\varepsilon v\|_{BMO} \leq \|\partial_x^{-1}[\nu - P_\varepsilon \nu]\|_\infty = \max_J |\langle \chi_J - \chi_J * P_\varepsilon, \nu \rangle|$$
$$\lesssim \max \nu(I)$$

where $J \subset \mathbf{R}$ (resp. $I \subset \mathbf{R}$) refers to an interval (resp. an interval of size $\varepsilon$). Recall that $\nu$ is positive. Take smooth $\tau$

$$0 \leq \tau \leq 1, \quad \int \tau \sim \varepsilon, \quad |\tau'| \lesssim \frac{1}{\varepsilon}$$

Thus

$$\nu(I) \leq \langle \tau, \nu \rangle$$
$$\leq |\langle \partial_x^{-1} \mathcal{H}[\nu], \partial_x \mathcal{H} \tau \rangle|$$
$$= |\langle v, \partial_x \mathcal{H} \tau \rangle|$$
$$\leq |\langle u_0, \partial_x \mathcal{H} \tau \rangle| + |\langle u_1, \partial_x \mathcal{H} \tau \rangle| + |\langle h, \partial_x \mathcal{H} \tau \rangle|$$
$$\leq \|u_0\|_\infty \|\partial_x \mathcal{H} \tau\|_1 + \|u_1\|_1 \|\partial_x \mathcal{H} \tau\|_\infty + \|\partial_x \mathcal{H} h\|_\infty \|\tau\|_1$$
$$\leq C\varepsilon_0 + C\frac{\varepsilon_1}{\varepsilon} + C\varepsilon B$$

(observe that $\partial_x \tau$ in an "atom").

From choice of $\varepsilon$,

$$\nu(I) < C(\varepsilon_0 + \sqrt{\varepsilon_1 B})$$

for $I$ as above and also

$$\|u\|_{BMO} < C(\varepsilon_0 + \sqrt{\varepsilon_1 B})$$

This proves the inequality.

**Corollary 4.10.** *Assume $u$ as above, satisfying*

$$\text{mes}[x \in \mathbf{T} \mid |u(x) - \langle u \rangle| > \varepsilon_0] < \varepsilon_1$$

*Then*

$$\text{mes}[x \in \mathbf{T} \mid |u(x) - \langle u \rangle| > \sqrt{\varepsilon_0}] < e^{-c[\sqrt{\varepsilon_0} + \sqrt{\frac{\varepsilon_1 B}{\varepsilon_0}}]^{-1}} \tag{4.11}$$

**Proof.** From (4.9) and the John-Nirenberg inequality.

We also will need estimates on functions of several variables with pluri-subharmonic extension. We only treat the case of two variables in Lemma 4.12 below. Similar inequalities for $d$ variables are obtained along the same lines.

Thus, in the remainder of this chapter, assume that $u = u(x, y)$ is 1-periodic in $x, y$, $|u| \leq 1$, and with pluri-subharmonic extension, $\tilde{u} = \tilde{u}(\tilde{x}, \tilde{y})$, $|\text{Im} \tilde{x}| \leq 1$, $|\text{Im} \tilde{y}| \leq 1$ satisfying

$$|\tilde{u}| < B$$

**Lemma 4.12.** *Let $u$ be as above, satisfying, moreover,*

$$\text{mes}[(x, y) \in \mathbf{T}^2 \big| |u(x, y) - \langle u \rangle| > \varepsilon_0] < \varepsilon_1$$

*Then*

$$\text{mes}[(x, y) \in \mathbf{T}^2 \big| |u - \langle u \rangle| > \varepsilon_0^{1/4}] < \exp \left\{ -c \left( \varepsilon_0^{1/4} + \left( \frac{B}{\varepsilon_0} \right)^{1/2} \varepsilon_1^{1/4} \right)^{-1} \right\}$$

$$(4.13)$$

**Proof.** We may assume that $\langle u \rangle = 0$. Write $u = u_0 + u_1$, $\|u_0\|_\infty < \varepsilon_0$, $\|u_1\|_1 < \varepsilon_1$.
Denote

$$A = \{y \in \mathbf{T} \big| \int |u_1(x, y)| dx < \sqrt{\varepsilon_1}\}$$

Hence

$$\text{mes} \, (\mathbf{T} \backslash A) < \sqrt{\varepsilon_1}$$

Denote

$$U(y) = \int_{\mathbf{T}} u(x, y) dx$$

Thus, for $y \in A$, we have

$$|U(y)| < \varepsilon_0 + \sqrt{\varepsilon_1}$$

and from the one-variable result applied in $x$, we have

$$\|u(\cdot, y)\|_{BMO} \leq C(\varepsilon_0 + \varepsilon_1^{1/4} B^{1/2})$$

and hence

$$\int e^{c \frac{|u(x,y)|}{\varepsilon_0 + \varepsilon_1^{1/4} B^{1/2}}} dx < C$$

Integrating in $y \in A$,

$$\int_A \int e^{c \frac{|u(x,y)|}{\varepsilon_0 + \varepsilon_1^{1/4} B^{1/2}}} dx dy < C$$

Denote

$$\gamma = e^{-c \frac{\varepsilon_0^{1/2}}{\varepsilon_0 + \varepsilon_1^{1/4} B^{1/2}}}$$

$$B = \{x \in \mathbf{T} \big| \int_A e^{c\frac{|u(x,y)|}{\varepsilon_0 + \varepsilon_1^{1/4} B^{1/2}}} \, dy < \gamma^{-1}\}$$

Thus

$$\mathrm{mes}\,(\mathbf{T}\backslash B) < C\gamma$$

For $x \in B$, write

$$\mathrm{mes}\,[y \in \mathbf{T} \big| \, |u(x,y)| > 2\varepsilon_0^{1/2}] <$$

$$\mathrm{mes}\, A^c + \gamma^{-1} e^{-2c\frac{\varepsilon_0^{1/2}}{\varepsilon_0 + \varepsilon_1^{1/4} B^{1/2}}} <$$

$$\sqrt{\varepsilon_1} + e^{-c\frac{\varepsilon_0^{1/2}}{\varepsilon_0 + \varepsilon_1^{1/4} B^{1/2}}}$$

We may clearly assume that $e^{-c\frac{\varepsilon_0^{1/2}}{\varepsilon_0 + \varepsilon_1^{1/4} B^{1/4}}} < \varepsilon_1$ since (4.13) is otherwise obvious.
For fixed $x \in B$, apply the one-variable result in $y$. Hence

$$\mathrm{mes}\,[y \big| \, |u(x,y)| > \varepsilon_0^{1/4}] < e^{-c[\varepsilon_0^{1/4} + (\frac{\varepsilon_1}{\varepsilon_0})^{1/4} B^{1/2}]^{-1}}$$

Therefore,

$$\mathrm{mes}\,[|u| > \varepsilon_0^{1/4}] < \mathrm{mes}\,(\mathbf{T}\backslash B) + e^{-c[\varepsilon_0^{1/4} + (\frac{\varepsilon_1}{\varepsilon_0})^{1/4} B^{1/2}]^{-1}}$$

$$< e^{-c[\varepsilon_0^{1/2} + \varepsilon_0^{-1/2}\varepsilon_1^{1/4} B^{1/2}]^{-1}} + e^{-c[\varepsilon_0^{1/4} + (\frac{\varepsilon_1}{\varepsilon_0})^{1/4} B^{1/2}]^{-1}}$$

$$< e^{-c[\varepsilon_0^{1/4} + (\frac{B}{\varepsilon_0})^{1/2}\varepsilon_1^{1/4}]^{-1}}$$

This proves (4.13).

**Remark.** In the case of $d$ variables, exponents $\frac{1}{4}$ in (4.13) should be replaced by $\frac{1}{2d}$.

## Reference

[B-G-S] J. Bourgain, M. Goldstein, W. Schlag. Anderson localization for Schrödinger operators on $\mathbf{Z}^b$ with potentials given by the skew shift, *CMP* 220(3) (2002), 583–621.

# Chapter Five

## LDT for Shift Model

This chapter is related primarily to [B-G] and [G-S] (see Chapter 1).

For $T$ given by a shift $Tx = x + \omega$, with $\omega$ satisfying a diophantine condition, we establish inequalities of the form (2.6) for $\frac{1}{N} \log \|M_N(x, E)\|$. We will use the results from the previous chapter on subharmonic functions. We first treat the case $d = 1$ and then $d = 2$.

**Theorem 5.1.** *Assume that* $\omega \in \mathbf{T}$ *satisfies a DC*

$$\|k\omega\| > c\frac{1}{|k|\left(\log(1 + |k|)\right)^3} \quad \text{for } k \in \mathbf{Z}\backslash\{0\} \tag{5.2}$$

*Let* $v$ *be real analytic on* $\mathbf{T}$ *and*

$$M_N(x) = \prod_N^1 \left( \begin{array}{cc} v(x + j\omega) - E & -1 \\ 1 & 0 \end{array} \right)$$

*Then, for* $\kappa > N^{-1/10}$,

$$\text{mes}[x \in \mathbf{T} \mid \left| \frac{1}{N} \log \|M_N(x)\| - L_N(E) \right| > \kappa] < Ce^{-c\kappa^2 N} \tag{5.3}$$

**Remark.** If (5.2) is weakened to a DC

$$\|k\omega\| > c|k|^{-A} \text{ for } k \in \mathbf{Z}\backslash\{0\}$$

we still get a conclusion

$$\text{mes}[x \in \mathbf{T} \mid \left| \frac{1}{N} \log \|M_N(x)\| - L_N(E) \right| > N^{-\sigma}] < Ce^{-N^\sigma} \tag{5.4}$$

for some $\sigma = \sigma(A) > 0$.

**Proof of Theorem 5.1.** By assumption, $M_N(x)$ has an analytic extension $M_N(z)$ to a strip $|\text{Im } z| < \rho$, for some $\rho > 0$, satisfying $\|M_N(z)\| < C^N$. Hence

$$u(x) = \frac{1}{N} \log \|M_N(x)\|$$

has bounded subharmonic extension to $|\text{Im } z| < \rho$, and therefore,

$$|\hat{u}(k)| \lesssim \frac{1}{|k|}$$

It is also clear from the definition of $u$ and $M_N(x)$ that

$$|u(x) - u(x + \omega)| < \frac{C}{N}$$

We use here that $\begin{pmatrix} v(x+j\omega) - E & -1 \\ 1 & 0 \end{pmatrix} \in SL_2(\mathbf{R})$ are bounded with bounded inverse.

Expand $u$ as a Fourier series

$$u(x) = L_N(E) + \sum_{k\in\mathbf{Z}\backslash\{0\}} \hat{u}(k)e^{2\pi ikx}$$

By assumption on $\omega$, we may find an approximant $q$ of $\omega$ satisfying

$$N < q < N(\log N)^3$$

Thus

$$\|j\omega\| > \frac{1}{2q} \text{ if } 1 \leq j < q$$

Write

$$u(x) = \sum_{|j|<R} \frac{R-|j|}{R^2} u(x+j\omega) + 0\left(\frac{R}{N}\right) \qquad (R \sim \kappa N)$$

$$u(x) - L_N(E) = \sum_{k\neq 0} \hat{u}(k)\left[ \sum_{|j|<R} \frac{R-|j|}{R^2} e^{2\pi ijk\omega} \right]e^{2\pi ikx} + 0\left(\frac{R}{N}\right)$$

where

$$\left| \sum_{|j|<R} \frac{R-|j|}{R^2} e^{2\pi ijk\omega} \right| < \frac{1}{1+R^2\|k\omega\|^2}$$

Split the sum

$$\sum_{k\neq 0} = \sum_{0<|k|<N^{1/2}} + \sum_{N^{1/2}\leq|k|<q} + \sum_{q\leq|k|<K} + \sum_{|k|>K}$$

where

$$\log K \sim \kappa^2 N$$

The first three sums are bounded uniformly in $x$. The last is bounded in the mean

$$(I) < \sum_{0<|k|<N^{1/2}} |\hat{u}(k)| \frac{1}{1+R^2\|k\omega\|^2}$$

$$< \sum_{0<|k|<N^{1/2}} \frac{1}{|k|} \frac{1}{1+R^2\|k\omega\|^2}$$

$$< \sum_{0<|k|<N^{1/2}} \frac{1}{R|k|\,\|k\omega\|} < \frac{(\log N)^3 N^{1/2}}{R} \lesssim \frac{(\log N)^3}{\kappa N^{1/2}} < \kappa^{-1/2}$$

$$(II) < \sum_{N^{1/2}\leq|k|<q} \frac{1}{|k|} \frac{1}{1+R^2\|k\omega\|^2} < N^{-1/2} \sum_{0<|k|<q} \frac{1}{1+R^2\|k\omega\|^2}$$

Let $I \subset \mathbf{Z}$ be an interval of size $|I| < q$. From assumption on $q$, we have that $\|(k_1 - k_2)\omega\| > \frac{1}{2q}$ for $k_1 \neq k_2$ in $I$ and therefore

$$\sum_{k\in I} \frac{1}{1+R^2\|k\omega\|^2} \lesssim 1 + \sum_{1\leq j\leq q} \frac{1}{1+R^2\frac{j^2}{q^2}} \lesssim \frac{q}{R}$$

Hence

$$(II) < N^{-1/2} \frac{q}{R} < N^{-1/2} \frac{N(\log N)^3}{\kappa N} < \kappa$$

Similarly,

$$(III) < \sum_{q \leq |k| < K} \frac{1}{|k|} \frac{1}{1 + R^2 \|k\omega\|^2}$$

and subdividing $[q, K]$ in intervals of size $q$, we get

$$(III) < \sum_{1 \leq s < K/q} \frac{1}{s.q} \cdot \frac{q}{R} < \frac{1}{R} \log \frac{K}{q} < \frac{\kappa^2 N}{\kappa N} < \kappa$$

Finally, since

$$(IV) = \sum_{|k| > K} \hat{u}(k) \left[ \sum_{|j| < R} \frac{R - |j|}{R^2} e^{2\pi i j k \omega} \right] e^{2\pi i k x}$$

we obtain that

$$\int |(IV)|^2 dx \leq \sum_{|k| > K} |\hat{u}(k)|^2 \leq \sum_{|k| > K} \frac{1}{k^2} < K^{-1}$$

$$< e^{-c\kappa^2 N}$$

This proves (5.3).

Consider next the case of a shift $Tx = x + \omega$ on $\mathbf{T}^d$ for $d > 1$. Take $d = 2$ (the general case is similar, using the analogue of Lemma 4.12 in $d$ variables).

**Theorem 5.5.** *Assume that $\omega \in \mathbf{T}^2$ satisfies a DC*

$$\|k\omega\| > c|k|^{-A} \quad \text{for } k \in \mathbf{Z}^2 \backslash \{0\}$$

*Let $v$ be real analytic on $\mathbf{T}^2$. Then there is $\sigma = \sigma(A) > 0$ s.t.*

$$mes \left[ x \in \mathbf{T}^2 \,\Big|\, \left| \frac{1}{N} \log \|M_N(x, E)\| - L_N(E) \right| > N^{-\sigma} \right] < e^{-N^\sigma}$$

There does not seem to be an immediate extension of the previous argument for $d > 1$. We will first obtain a weaker inequality that will then be improved using Lemma 4.12.

**Proof of Theorem 5.5.** Denote $x = (x_1, x_2) \in \mathbf{T}^2$. Fixing one variable, we clearly obtain

$$|\widehat{u(x_1)}(k_2)| < \frac{C}{k_2} \quad \text{and} \quad |\widehat{u(x_2)}(k_1)| < \frac{C}{|k_1|}$$

and hence

$$\sum_{k_1} |\hat{u}(k_1, k_2)|^2 < \frac{C}{k_2^2} \quad \text{and} \quad \sum_{k_2} |\hat{u}(k_1, k_2)|^2 < \frac{C}{k_1^2}$$

Again,

$$|u(x) - u(x + \omega)| < \frac{C}{N}$$

Estimate

$$\|u - L_N(E)\|_2 \le \left\| \frac{1}{R} \sum_{j=1}^{R} u(x + j\omega) - \hat{u}(0) \right\|_2 + 0\left(\frac{R}{N}\right)$$

$$< \left[ \sum_{k \in \mathbf{Z}^2 \setminus \{0\}} |\hat{u}(k)|^2 \frac{1}{1 + R^2 \|k\omega\|^2} \right]^{1/2} + 0\left(\frac{R}{N}\right)$$

Fix some $K$ and split the sum in $\sum_{0 < |k| < K} + \sum_{|k| > K} = (I) + (II)$.
Estimate from assumption on $\omega$ and preceding

$$(I) < \left[ \sum_{0 < |k| < K} \frac{1}{|k|^2} \frac{1}{(R|k|^{-A})^2} \right] < \frac{K^{2A}}{R^2}$$

$$(II) < \sum_{|k| > K} |\hat{u}(k)|^2 < \frac{C}{K}$$

Hence

$$\|u - L_N(E)\|_2 \lesssim \frac{K^A}{R} + \frac{C}{K^{1/2}} + \frac{R}{N} < N^{-\rho}$$

for appropriate choice of $R$ and $K$ and some $\rho > 0$.
Thus letting $\varepsilon_0 = N^{-\frac{2\rho}{5}}$, $\varepsilon_1 = N^{-\frac{6\rho}{5}}$, we have

$$\text{mes}\left[ x \in \mathbf{T}^2 \middle| \, |u - \langle u \rangle| > \varepsilon_0 \right] < \varepsilon_1$$

Invoking Lemma 4.12, this implies that

$$\text{mes}\left[ x \in \mathbf{T}^2 \middle| \, |u - \langle u \rangle| > \varepsilon_0^{1/4} \right] < \exp\{-c(\varepsilon_0^{1/4} + \varepsilon_0^{-1/2}\varepsilon_1^{1/4})^{-1}\}$$

and hence

$$\text{mes}\left[ x \in \mathbf{T}^2 \middle| \, |u - L_N(E)| > N^{-\rho/10} \right] < e^{-cN^{\rho/10}}$$

This proves Theorem 5.5 with $\sigma = \rho/10$.

**Problem.** Does for typical $\omega \in \mathbf{T}^2$ and fixed $\kappa > 0$

$$\text{mes}\left[ x \in \mathbf{T}^2 \middle| \, \left| \frac{1}{N} \log \|M_N(x, E)\| - L_N(E) \right| > \kappa \right] < e^{-\varepsilon(\kappa)N}$$

hold?

# Chapter Six

## Avalanche Principle in $SL_2(\mathbf{R})$

The main result of this chapter is Proposition 6.1 from [G-S] (see Chapter 1).

**Proposition 6.1.** *Let* $A_1, \cdots, A_n$ *be a sequence in* $SL_2(\mathbf{R})$ *satisfying the conditions*

$$\min_{1 \leq j \leq n} \|A_j\| \geq \mu > n \tag{6.2}$$

$$\max_{1 \leq j \leq n} |\log \|A_j\| + \log \|A_{j+1}\| - \log \|A_{j+1}A_j\|| < \frac{1}{2} \log \mu \tag{6.3}$$

*Then*

$$|\log \|A_n \cdots A_1\| + \sum_{j=2}^{n-1} \log \|A_j\| - \sum_{j=1}^{n-1} \log \|A_{j+1}A_j\|| < C \frac{n}{\mu} \tag{6.4}$$

Some notation used in the proof:

Assume $K \in SL_2(\mathbf{R})$. Denote $u_K^{\pm}$ the normalized eigenvectors of $\sqrt{K^*K}$. Thus

$$Ku_K^+ = \|K\|v_K^+ \qquad Ku_K^- = \|K\|^{-1}v_K^-$$

where

$$\|v_K^+\| = 1 = \|v_K^-\|$$

Given $K, M \in SL_2(\mathbf{R})$, denote further for $\varepsilon_1 = \pm 1, \varepsilon_2 = \pm 1$

$$b^{\varepsilon_1, \varepsilon_2}(K, M) = v_K^{\varepsilon_1} \cdot u_M^{\varepsilon_2}$$

(only defined up to the sign).

**Proof of Proposition 6.1.** First, we observe that

$$\|MK\| \geq \|MKu_K^+\| = \|K\| \, \|Mv_K^+\|$$
$$\geq \|K\|(b^{+,+}(K, M) . \|M\| - \|M\|^{-1})$$

and also

$$\|MK\| \leq b^{+,+}(K, M)\|K\| . \|M\| + \|K\|^{-1}\|M\| + \|K\| \, \|M\|^{-1}$$

In particular

$$\frac{\|A_{j+1}A_j\|}{\|A_j\| \, \|A_{j+1}\|} + \frac{1}{\|A_{j+1}\|^2} \geq b^{+,+}(A_j, A_{j+1}) \geq \frac{\|A_{j+1}A_j\|}{\|A_j\| \, \|A_{j+1}\|} - \frac{1}{\|A_j\|^2} - \frac{1}{\|A_{j+1}\|^2} \tag{6.5}$$

Next, one gets for any vector $u$

$$A_n \cdots A_1 u = \sum_{\varepsilon_1,\ldots,\varepsilon_n = \pm 1} \|A_n\|^{\varepsilon_n} \left[ \prod_{j=1}^{n-1} \|A_j\|^{\varepsilon_j} \, b^{\varepsilon_j, \varepsilon_{j+1}}(A_j, A_{j+1}) \right] \langle u_{A_1}^{\varepsilon_1}, u \rangle v_{A_n}^{\varepsilon_n}$$

(6.6)

It follows from (6.5), (6.2), and (6.3) that

$$|b^{+,+}(A_j, A_{j+1})| \frac{\|A_j\| \, \|A_{j+1}\|}{\|A_{j+1} A_j\|} = 1 + 0\left(\frac{\mu^{1/2}}{\mu^2}\right)$$

$$|b^{+,+}(A_j, A_{j+1})| > \mu^{-1/2} - 0(\mu^{-2}) > \frac{1}{2\sqrt{\mu}}$$

By (6.6),

$$\|A_n \cdots A_1\| \le \sum_{\varepsilon_1,\ldots,\varepsilon_n = \pm 1} \|A_n\|^{\varepsilon_n} \left[ \prod_{j=1}^{n-1} \|A_j\|^{\varepsilon_j} \, |b^{\varepsilon_j, \varepsilon_{j+1}}(A_j, A_{j+1})| \right]$$

and also, evaluating at $u = u_{A_1}^+$,

$$\|A_n \cdots A_1\| \ge \|A_n\| \prod_{j=1}^{n-1} \|A_j\| \, |b^{+,+}(A_j, A_{j+1})| - \sum_{\substack{\varepsilon_1,\ldots,\varepsilon_n = \pm 1 \\ \min \varepsilon_j = -1}} \|A_n\|^{\varepsilon_n} \prod_{j=1}^{n-1} \|A_j\|^{\varepsilon_j} \, |b^{\varepsilon_j, \varepsilon_{j+1}}(A_j, A_{j+1})|$$

(6.7)

The second term in (6.7) may be bounded by

$$\|A_n\| \left[ \prod_{j=1}^{n-1} \|A_j\| \, |b^{+,+}(A_j, A_{j+1})| \right] \gamma$$

where

$$\gamma \le \sum_{\substack{\varepsilon_1,\ldots,\varepsilon_n = \pm 1 \\ \min \varepsilon_j = -1}} \left[ \prod_{j=1}^{n} \|A_j\|^{\varepsilon_j - 1} \right] \left[ \prod_{j=1}^{n-1} \frac{|b^{\varepsilon_j, \varepsilon_{j+1}}(A_j, A_{j+1})|}{|b^{+,+}(A_j, A_{j+1})|} \right]$$

Denoting $\ell = \#\{j = 1, \ldots, n | \varepsilon_j = -1\}$, this gives

$$\gamma \le \sum_{\ell=1}^{n} \binom{n}{\ell} \mu^{-2\ell} (2\sqrt{\mu})^{2\ell} < \left(1 + \frac{4}{\mu}\right)^n - 1 < C\frac{n}{\mu}$$

Thus we proved that

$$\frac{\|A_n\| \prod_{j=1}^{n-1} \|A_j\| \, |b^{+,+}(A_j, A_{j+1})|}{\|A_n \cdots A_1\|} = 1 + 0\left(\frac{n}{\mu}\right)$$

$$\left| \log \|A_n \cdots A_1\| - \sum_{j=1}^{n} \log \|A_j\| - \sum_{j=1}^{n-1} \log |b^{+,+}(A_j, A_{j+1})| \right| < C\frac{n}{\mu} \quad (6.8)$$

Since also for each $j$

$$\left| \log |b^{+,+}(A_j, A_{j+1})| - \log \|A_{j+1} A_j\| + \log \|A_j\| + \log \|A_{j+1}\| \right| < c\mu^{-3/2}$$

(6.8) implies further that

$$\left| \log \|A_n \cdots A_1\| + \sum_{2}^{n-1} \log \|A_j\| - \sum_{1}^{n-1} \log \|A_{j+1} A_j\| \right| < C\frac{n}{\mu}$$

This proves Proposition 6.1.

# Chapter Seven

---

## Consequences for Lyapounov Exponent, IDS, and Green's Function

The results from this chapter appear in [B-G] and [G-S] (see Chapter 1).

The rotation vector $\omega$ is assumed to satisfy a diophantine condition (DC).

Results for general $\omega$ have been obtained more recently in [B-J] and [B] but will not be presented here.

Again, let the transformation $T$ be given by the $\omega$-shift on $\mathbf{T}^d$, $d \geq 1$, assuming $\omega$ satisfying a DC. The potential $v = v(x)$ is assumed real analytic on $\mathbf{T}^d$.

Recall the LDT from Chapter 5 for

$$M_N(E, x) = \prod_N^1 \begin{pmatrix} v(x + j\omega) - E & -1 \\ 1 & 0 \end{pmatrix}$$

Thus, fixing a small $\kappa > 0$ and taking $N$ large enough, we have

$$\text{mes}\,[x \in \mathbf{T}^d| \,\left|\frac{1}{N} \log \|M_N(E, x)\| - L_N(E)\right| > \kappa] < e^{-cN^\sigma} \tag{7.1}$$

for some constants $c, \sigma > 0$.

For $d = 1$ and $\omega$ typical, we may take $\sigma = 1$. We will combine (7.1) with the result from the previous chapter. First, we prove the positivity of $L(E)$ in the perturbative regime.

**Proposition 7.2.** *Let $v_0$ be nonconstant real analytic on $\mathbf{T}^d$ and $v = \lambda v_0$ with $\lambda > \lambda_0(v, \omega)$. Then, for all energies $E$,*

$$L(E) > \frac{1}{2} \log \lambda$$

Thus, compared with Proposition 3.3, Proposition 7.2 holds in any dimension, but the proof below requires a DC on $\omega$. See also the comments at the end of this chapter.

We will use the following consequence of the classical Lojasiewicz result.

**Lemma 7.3.** *There is a constant $c_0 = c_0(v_0)$ s.t.*

$$\text{mes}[x \in \mathbf{T}^d| \,|v_0(x) - E_1| < \delta] < \delta^{c_0} \tag{7.4}$$

*for all $E_1$ and sufficiently small $\delta > 0$.*

**Proof of Proposition 7.2.** Fix a large scale $n_0$ and chose

$$\delta = n_0^{-\frac{2}{c_0}} \quad \text{and} \quad \lambda = \delta^{-10}$$

Choice of $n_0$ depends on diophantine assumptions on $\omega$.

If $v = \lambda v_0$, it follows from (7.4) that for $n \leq n_0$ and $E$ arbitrary,

$$\|M_n(E, x)\| > (\delta\lambda - 1)^n$$

for $x$ outside a set of measure at most $n\delta^{c_0} < \frac{1}{n_0}$. Hence,

$$\log(\lambda + |E|) + C > L_n(E) = \frac{1}{n}\int \log\|M_n(E, x)\|dx > \left(1 - \frac{1}{n_0}\right)\log(\delta\lambda - 1)$$
$$> \frac{4}{5}\log\lambda$$

Fix $E$, $|E| < C\lambda$ (otherwise there is nothing to prove).

Using the submultiplicity $L_{2n}(E) \leq L_n(E)$, the preceding permits us to find $n_1 < n_0$, $n_1 \sim n_0$ satisfying

$$|L_{2n_1}(E) - L_{n_1}(E)| < \frac{1}{100}L_{n_1}(E) \tag{7.5}$$

Observe that the functions $\frac{1}{N}\log\|M_N(E, x)\|$ are bounded by $\log\lambda + C$ as well as their subharmonic extension. Since $\lambda$ is large, the statement of the LDT requires first proper normalization. Thus we let

$$u(x) = \frac{1}{N\log\lambda}\cdot\log\|M_N(E, x)\|$$

and (7.1) needs to be restated as

$$\text{mes}\left[x \in \mathbf{T}^d\Big|\left|\frac{1}{N}\log\|M_N(x)\| - L_N(E)\right| > \kappa\log\lambda\right] < e^{-cN^\sigma}$$

Take $\kappa = 10^{-2}$. If

$$n_2 = mn_1 \sim e^{\frac{c}{2}n_1^\sigma}$$

the LDT permits us to ensure that

$$\max_{1\leq j\leq n_2}\left|\frac{1}{n_1}\log\|M_{n_1}(x + j\omega)\| - L_{n_1}(E)\right| < \frac{1}{50}L_{n_1}(E) \tag{7.6}$$

and

$$\max_{1\leq j\leq n_2}\left|\frac{1}{2n_1}\log\|M_{2n_1}(x + j\omega)\| - L_{2n_1}(E)\right| < \frac{1}{50}L_{2n_1}(E) \tag{7.7}$$

except for $x \in \Omega \subset \mathbf{T}^d$,

$$\text{mes}\,\Omega < e^{-\frac{c}{2}n_1^\sigma}$$

Fix $x \notin \Omega$ and define for $k = 1, \cdots, m$

$$A_k = M_{n_1}(x + (k-1)n_1\omega) \in SL_2(\mathbf{R})$$

We verify the conditions of Proposition 6.1.

By (7.6)

$$\|A_k\| > e^{n_1\frac{49}{50}L_{n_1}(E)} = \mu > e^{\frac{3}{4}(\log\lambda)n_1} > m$$

and by (7.6), (7.7), and (7.5)

$$\big|\log\|A_k\| + \log\|A_{k+1}\| - \log\|A_{k+1}A_k\|\big|$$
$$< |2n_1 L_{n_1}(E) - 2n_1 L_{2n_1}(E)| + \tfrac{1}{10}n_1 L_{n_1}(E)$$
$$< \tfrac{1}{5}n_1 L_{n_1}(E) < \tfrac{1}{2}\log\mu$$

Thus (6.2) and (6.3) hold, and (6.4) implies

$$\Big|\log\|A_m\cdots A_1\| + \sum_{2}^{m-1}\log\|A_k\| - \sum_{1}^{m-1}\log\|A_{k+1}A_k\|\Big| < C\frac{m}{\mu}$$

Hence, for $x$ outside $\Omega$,

$$\Big|\log\|M_{n_2}(x)\| + \sum_{2}^{m-1}\log\|M_{n_1}\big(x+(k-1)n_1\omega\big)\|$$
$$- \sum_{1}^{m-1}\log\|M_{2n_1}\big(x+(k-1)n_1\omega\big)\|\,\Big|$$
$$< e^{-\frac{1}{2}n_1(\log\lambda)}$$

Integration in $x$ thus implies

$$\Big|L_{n_2}(E) + \frac{m-2}{m}L_{n_1}(E) - 2\frac{m-1}{m}L_{2n_1}(E)\Big| < e^{-\frac{n_1}{2}\log\lambda} + 2L_{n_1}(E).\mathrm{mes}\,\Omega$$
$$< 3\log\lambda\, e^{-\frac{c}{2}n_1^\sigma} \qquad (7.8)$$

$$\big|L_{n_2}(E) + L_{n_1}(E) - 2L_{2n_1}(E)\big| < \frac{L_{n_1}(E)}{m} + 3\log\lambda\, e^{-\frac{c}{3}n_1^\sigma} < \frac{5n_1}{n_2}\log\lambda$$
$$< e^{-\frac{c}{3}n_1^\sigma} \qquad (7.9)$$

Writing the same inequality with $n_2$ replaced by $2n_2$ and subtracting gives

$$\big|L_{2n_2}(E) - L_{n_2}(E)\big| < 2e^{-\frac{c}{3}n_1^\sigma} \qquad (7.10)$$

Also, again recalling (7.5),

$$\big|L_{n_2}(E) - L_{n_1}(E)\big| \le 2|L_{2n_1}(E) - L_{n_1}(E)| + e^{-\frac{c}{3}n_1^\sigma} < \frac{1}{40}L_{n_1}(E) \qquad (7.11)$$

The iteration is now clear.

At the next step, replace $n_1$ by $n_2$ and let $n_3 \sim e^{\frac{c}{2}n_2^\sigma}$, etc. We get

$$|L(E) - L_{n_1}(E)$$
$$< \sum_{s\ge1}|L_{n_{s+1}}(E) - L_{n_s}(E)|$$
$$< 2\sum_{s\ge1}\big(|L_{2n_s}(E) - L_{n_s}(E)| + e^{-\frac{c}{3}n_s^\sigma}\big)$$
$$< 2|L_{2n_1}(E) - L_{n_1}(E)| + 4\sum_{s\ge1}e^{-\frac{c}{3}n_s^\sigma} \qquad < \frac{1}{50}L_{n_1}(E) + e^{-\frac{c}{4}n_1^\sigma}$$
$$< \frac{1}{40}L_{n_1}(E)$$

and

$$L(E) > \frac{3}{4}L_{n_1}(E) > \frac{1}{2}\log\lambda$$

**Proposition 7.12.** *Assume $L(E) > \delta_0 =$ fixed constant. Then for all $n \in \mathbf{Z}_+$ large enough*

$$|L(E) + L_n(E) - 2L_{2n}(E)| < e^{-cn^\sigma} \tag{7.13}$$

*where $\sigma$ is the exponent in (7.1) and $c = c(\delta_0) > 0$.*

**Proof.** The preceding argument shows that if we take $n = n_1$ large enough (depending on $\delta_0$) and $\log n_2 \sim n^\sigma$, then

$$|L_{n_2}(E) + L_n(E) - 2L_{2n}(E)| < e^{-cn^\sigma}$$

(see (7.9)) and also

$$|L(E) - L_{n_2}(E)| \leq \sum_{s \geq 2} |L_{n_{s+1}}(E) - L_{n_s}(E)|$$

$$\leq 2 \sum_{s \geq 2} \left( |L_{2n_s}(E) - L_{n_s}(E)| + e^{-cn_s^\sigma} \right)$$

$$< 4 \sum_{s \geq 1} e^{-cn_s^\sigma} < e^{-cn^\sigma}$$

This proves (7.13).

**Corollary 7.14.** *Assume $L(E) > \delta_0$ for all $E \in [E_1, E_2]$. Then, for $E, E' \in [E_1, E_2]$,*

$$|L(E) - L(E')| < C \exp[-c(\log|E - E'|^{-1})^\sigma] \tag{7.15}$$

*In particular, if $d = 1$ (and $\omega$ typical),*

$$|L(E) - L(E')| < C|E - E'|^\kappa \tag{7.16}$$

*for some $\kappa = \kappa(\delta_0) > 0$.*

**Proof.** Obviously,

$$|L_n(E) - L_n(E')| < C^n|E - E'|$$

so that by (7.13)

$$|L(E) - L(E')| < C^n|E - E'| + e^{-cn^\sigma}$$

and (7.15) follows from choice $n \sim \log \frac{1}{|E-E'|}$.

**Corollary 7.17.** *Assume that $\omega$ satisfies a DC. Then $L(E)$ is a continuous function of $E \in \mathbf{R}$.*

**Proof.** We claim that if $L(E) > \delta$, then also $L(E') > \frac{\delta}{2}$ for $|E - E'| < \varepsilon(\delta)$. This will in particular imply Corollary 7.17.

Choose sufficiently large $n_1$ (depending in particular on $\delta$) to ensure that

$$|L_{2n_1}(E) - L_{n_1}(E)| < 10^{-3} L_{n_1}(E)$$

For $|E' - E| < C^{-n_1}\delta = \varepsilon(\delta)$, also

$$L_{n_1}(E') > \frac{9}{10}\delta \text{ and } |L_{2n_1}(E') - L_{n_1}(E')| < 10^{-2} L_{n_1}(E')$$

Following again the proof of Proposition 7.2, we get

$$|L(E') - L_{n_1}(E')| < 10^{-1}L_{n_1}(E') + e^{-cn_1^{\sigma}}$$

and hence

$$L(E') > \frac{3}{4}L_{n_1}(E') > \frac{\delta}{2}$$

proving the claim.

Relating the Lyapounov exponent $L(E)$ and the integrated density of states (IDS) $N(E)$, the following regularity property is obtained for shift models with typical $\omega$.

**Proposition 7.18.** *Assume $L(E) > \delta_0$ for $E \in [E_1, E_2]$.  For $d = 1$, the IDS is Holder continuous on $[E_1, E_2]$. Thus, for some $\kappa = \kappa(\delta_0) > 0$,*

$$|N(E) - N(E')| < C|E - E'|^{\kappa} \quad \text{for } E, E' \in [E_1, E_2]$$

*For $d \geq 2$, we have*

$$|N(E) - N(E')| < C \exp[-c\left(\log \frac{1}{|E - E'|}\right)^{\sigma}] \quad \text{for } E, E' \in [E_1, E_2]$$

**Proof.** Recalling the Thouless formula

$$L(E) = \int \log|E - E'| dN(E')$$

we see that $L(E)$ and $N(E)$ are related by the Hilbert transform. Thus the claim follows from Corollary 7.14.

**Remarks.**

**1.** Consider the Almost Mathieu operator

$$H = \lambda \cos(x + n\omega)\delta_{nn'} + \Delta$$

in perturbative regime (i.e., $\lambda$ large). It may then be shown that for typical $\omega$, we have

$$|N(E) - N(E')| < C_{\kappa}|E - E'|^{\kappa}$$

for all $\kappa < \frac{1}{2}$. The exponent $\kappa = \frac{1}{2}$ is optimal because of the presence of gaps in the spectrum.

**2.** Concerning Proposition 7.2, the author established more recently the full analogue of Proposition 3.3, thus the Sorets-Spencer theorem with $\lambda > \lambda_0(v)$ for $d > 1$ (see [Bo] in Chapter 3).

**Problem.** Does Proposition 7.18 require the positivity assumption of the $L(E)$? What happens in the (nonperturbative) Almost Mathieu model, say, at $\lambda = 2$? A (negative) result on this issue will be pointed out in the next chapter.

Returning to Chapter 2, we state the following important consequence of the LDT to Green's function estimates (in the shift model). Recall (2.7)

$$|G_N(E, x)(n_1, n_2)| < \frac{\|M_{n_1}(x, E)\| \cdot \|M_{N-n_2}(T^{n_2}x, E)\|}{|\det[H_N(x) - E]|}$$

and the subsequent discussion.

**Proposition 7.19.** *Assume $L(E) > \delta_0$. Then for $N > N_0(\delta_0)$, there is a set $\Omega \subset \mathbf{T}^d$ satisfying for some $\sigma > 0$*

$$\operatorname{mes}\Omega < e^{-cN^\sigma} \tag{7.20}$$

*and such that for any $x$ outside $\Omega$, one of the intervals*

$$\Lambda = [1, N]; [1, N-1]; [2, N]; [2, N-1]$$

*will satisfy*

$$|G_\Lambda(E, x)(n_1, n_2)| < e^{-L(E)|n_1 - n_2| + N^{1-}} \tag{7.21}$$

## References

[B] J. Bourgain. Positivity and continuity of the Lyapounov exponent for shifts on $\mathbf{T}^d$ with arbitrary frequency vector and real analytic potential, preprint 2002.

[B-J] J. Bourgain, S. Jitomirskaya. Continuity of the Lyapounov exponent for quasi-periodic operators with analytic potential, *J. Stat. Phys.* 108(5–6) (2002), 1203–1218.

# Chapter Eight

## Refinements

The purpose of this chapter is to analyze in more detail the estimates of Chapter 7 for the small Lyapounov exponent $L(E)$. We consider only the case of the 1-frequency shift model

$$H(x) = v(x + n_\omega)\delta_{nn'} + \Delta \qquad (x, \omega \in \mathbf{T}) \tag{8.1}$$

with $v$ 1-periodic and with bounded analytic extension on $z = x + iy, |y| \le 1$, say. Assume again the rotation number $\omega$ satisfying (5.2)

$$\|k\omega\| > c\frac{1}{|k|[\log(1 + |k|)]^3} \text{ for } k \in \mathbf{Z}\backslash\{0\} \tag{8.2}$$

(this assumption may be replaced by weaker ones).

**Proposition 8.3.** *Assume that the Lyapounov exponent $L(\cdot)$ of (8.1) satisfi es*

$$L(E) > 0 \text{ for } E \in [E_1, E_2] \subset \mathbf{R}$$

*Then $L(\cdot)$ and the IDS $N(\cdot)$ are Holder continuous on $[E_1, E_2]$ with exponent $c > 0$ depending only on the bound for the analytic extension of $v$ $\big($assuming $\omega$ satisfi es (8.2)$\big)$.*

Without making any positivity assumption on $L(\cdot)$, we may state

**Proposition 8.4.** *For any $A > 0$, we have the unconditional estimate*

$$|N(E) - N(E')| \le C_A \left(\log \frac{1}{|E - E'|}\right)^{-A} \tag{8.5}$$

*for $|E - E'| < \frac{1}{2}$.*

Proposition 8.4 improves on the general log-Hölder regularity result for stochastic Jacobi matrices (see [C-S]). There is the following immediate corollary of Proposition 8.3 for the Almost Mathieu operator.

**Corollary 8.6.** *For $\lambda > 2$, the IDS of*

$$H(x) = \lambda \cos(x + n\omega)\delta_{nn'} + \Delta \tag{8.7}$$

*is Hölder continuous with exponent $c > 0$ independent of $\lambda$.*

**Remark.** The constant $K$ in the inequality

$$|N(E) - N(E')| < K|E - E'|^c$$

does depend on the lower bound on $L(E)$, however. Otherwise, the IDS of (8.7) at $\lambda = 2$ would be Hölder continuous. Now let

$$\omega = \cfrac{1}{a_1 + \cfrac{1}{a_2 + \cfrac{1}{a_3 + \cfrac{1}{\ddots}}}}$$

be the continuous fraction expansion of $\omega$, and assume that all convergents satisfy

$$|a_s| > C(\varepsilon) \tag{8.8}$$

where $C(\varepsilon)$ is a sufficiently large constant depending on given $\varepsilon > 0$. According to the results from Helffer-Sjostrand [H-S], one may then write

$$\operatorname{Spec} H_{\lambda=2} \subset I_0 \cup \bigcup_{1 \leq j \leq a_1'} I_j \text{ with } a_1' \sim a_1$$

where $\{I_j\}$ are intervals, $|I_0| < \varepsilon$ and $|I_j| < e^{-\frac{1}{2}a_1}$ for $j \geq 1$. Furthermore, there is a renormalization of $H$ on each $I_j$, $j \geq 1$ with a similar description of the spectrum, $a_2$ replacing $a_1$, etc.

If $\varepsilon > 0$ is small enough, $\operatorname{Var} N(I_0) < \frac{1}{2}$, and hence $\operatorname{Var} N(I_j) > \frac{1}{2a_1'}$ for some $j \geq 1$. Iterating, one obtains after $s$ steps an interval $I$ satisfying

$$|I| < e^{-\frac{1}{2}(a_1 + \cdots + a_s)}$$

while

$$\operatorname{Var} N(I) > 2^{-s} \frac{1}{a_1' \cdots a_s'}$$

If thus $\frac{1}{s}(a_1 + \cdots + a_s) \to \infty$, which is typically the case, $N$ clearly cannot be Hölder regular.

We will first prove Proposition 8.3, and a more careful examination of the arguments will then lead to Proposition 8.4.

In the next few lemmas we establish upper bounds on the transfer matrix $\|M_N(x, E)\|$, valid for all $x \in \mathbf{T}$.

**Lemma 8.9.** *One has for all $x \in \mathbf{T}$ the upper bound*

$$\frac{1}{R} \sum_{j=0}^{R-1} \left( \frac{1}{N} \log \|M_N(x + j\omega; E)\| \right) < L_N(E) + c \frac{(\log R)^5}{R} \tag{8.10}$$

**Proof.** Denoting

$$u(x) = \frac{1}{N} \log \|M_N(x, E)\|$$

recall the Riesz-decomposition (4.2)

$$u(x) = \int \log |x - w| \mu(dw) + h(x) \quad \text{for } x \in [-1, 1]$$

with $\mu \geq 0$ and $h$ smooth. Fix $\delta > 0$ and majorize $u$ by the periodic function

$$u_{(\delta)}(x) = \sum_{j \in \mathbf{Z}} v_\delta(x + j)\eta(x + j) \tag{8.11}$$

with $v_{(\delta)}$ on $[-1, 1]$ defined by

$$v_{(\delta)}(x) = \int \log(|x - w| + \delta)\mu(dw) + h(x)$$

and $0 \leq \eta \leq 1$ a smooth function satisfying

$$\operatorname{supp}\eta \subset \, ] -1, 1[ \ \text{ and } \ \sum \eta(x+j) = 1$$

Thus clearly

$$u \leq u_{(\delta)} \text{ pointwise}$$

Since for $k \in \mathbf{Z}$

$$\widehat{u_{(\delta)}}(k) = \int_{\mathbf{R}} v_{(\delta)}(x)\eta(x)e^{-2\pi i k x}\,dx \equiv \widehat{v_\delta \eta}(k)$$

we obtain

$$\begin{aligned}
\widehat{u_{(\delta)}}(0) &= \widehat{u\eta}(0) + 0(\|(u-v_\delta)\eta\|_1)\\
&= \widehat{u}(0) + 0(\delta . \log \tfrac{1}{\delta})
\end{aligned} \tag{8.12}$$

and for $k \neq 0$

$$|\widehat{u}_{(\delta)}(k)| < C \min\left(\frac{1}{|k|}, \frac{1}{k^2\delta}\right) \tag{8.13}$$

The left side of (8.10) is bounded by

$$\frac{1}{R} \sum_{j=0}^{R-1} u_{(\delta)}(x+j\omega) =$$

$$\widehat{u}_\delta(0) + \sum_{k\neq 0} \widehat{u}_\delta(k)\left(\frac{1}{R}\sum_{j=0}^{R-1}e^{2\pi i k j \omega}\right)e^{2\pi i k x} = \quad (\text{by }(8.12))$$

$$L_N(E) + 0\left(\delta.\log\tfrac{1}{\delta} + \sum_{k\neq 0}|\widehat{u_\delta}(k)|\frac{1}{1+R\|k\omega\|}\right) \tag{8.14}$$

Let $\{q_s\}_{s\geq 1}$ be the approximants of $\omega$ satisfying by (8.2)

$$q_s < q_{s-1}(\log q_s)^3 \tag{8.15}$$

Fixing some $s_*$, estimate using (8.13)

$$\sum_{k\neq 0}|\widehat{u_\delta}(k)|\frac{1}{1+R\|k\omega\|} \leq$$

$$c\sum_{s\leq s_*}\sum_{q_{s-1}\leq |k|<q_s}\frac{1}{|k|R\|k\omega\|} + \frac{1}{\delta.q_{s_*}} \tag{8.16}$$

Since

$$\sum_{q_{s-1}<k<q_s}\frac{1}{k\|k\omega\|} < \frac{C}{q_{s-1}}\sum_{1\leq j<q_s}\frac{1}{j\|q_{s-1}\omega\|} < C\frac{q_s\log q_s}{q_{s-1}}$$

it follows from (8.15) that

$$(8.16) < Cs_*\frac{1}{R}(\log q_{s_*})^4 < C\frac{1}{R}(\log q_{s_*})^5 \tag{8.17}$$

Taking $\delta = \frac{1}{R}$ and $R^2 \leq q_{s_*} < R^2(\log R)^3$, substitution in (8.14) implies that

$$\frac{1}{R}\sum_{j=0}^{R-1}\left(\frac{1}{N}\log\|M_N(x+j\omega, E)\|\right) < L_N(E) + 0\left(\frac{1}{R}\log R + \frac{1}{R}(\log R)^5 + \frac{1}{R}\right)$$

and hence (8.10).

**Lemma 8.18.** $\frac{1}{N} \log \|M_N(x, E)\| < L_N(E) + C \frac{(\log N)^3}{N^{1/2}}$.

**Proof.** Estimate

$$\left| \log \|M_N(x + j\omega; E)\| - \log \|M_N(x, E)\| \right| \leq$$
$$\log \|M_j(x; E)\| + \log \|M_j(x + N\omega, E)\| < cj \qquad (8.19)$$

implying

$$\left| \frac{1}{N} \log \|M_N(x, E)\| - \frac{1}{R} \sum_{j=0}^{R-1} \left( \frac{1}{N} \log \|M_N(x + j\omega, E)\| \right) \right| < C \frac{R}{N}$$

Thus from Lemma 8.9

$$\frac{1}{N} \log \|M_N(x, E)\| < L_N(E) + C \frac{(\log R)^5}{R} + C \frac{R}{N}$$

and letting $R = N^{1/2}(\log N)^2$, (8.18) follows.

Returning to Theorem 5.1, the following improved LDT may be derived.

**Lemma 8.20.** *Assume*

$$L(E) > 0 \qquad (8.21)$$

*and fix $0 < \kappa < 1$. Then for $N$ sufficiently large*

$$\text{mes}\left[x \in \mathbf{T} \,\Big|\, \left| \frac{1}{N} \log \|M_N(x, E)\| - L(E) \right| > \kappa L(E) \right] < e^{-c\kappa^2 L(E).N} \qquad (8.22)$$

(The constant $c$ in the exponent does not depend on (8.21); inequality (5.3) gives the weaker inequality with $L(E)^2$ in the exponent – $L(E)$ is assumed small.)

**Proof.** Choose $N_0$ s.t.

$$N_0 > \kappa^{-2} \left( 1 + \frac{1}{L(E)^2} \right) \qquad (8.23)$$

$$L_{N_0}(E) < \left( 1 + \frac{\kappa}{2} \right) L(E) \qquad (8.24)$$

and let $N > N_0^3$.

Defining

$$u(x) = \frac{1}{N} \log \|M_N(x, E)\| \text{ and } R = \frac{\kappa}{10} N$$

write

$$|u(x) - \sum_{|j|<R} \frac{R-|j|}{R^2} u(x + j\omega)|$$
$$\leq \sum_{1 \leq j < R} \frac{R-j}{R^2} \left( \frac{1}{N} \log \|M_j(x, E)\| + \frac{1}{N} \log \|M_j(x + N\omega; E)\| \right) \qquad (8.25)$$

By (8.18),

$$(8.25) < \frac{2}{RN} \sum_{j=1}^{R} (jL_j(E) + j^{\frac{1}{2}+})$$

$$< \frac{2}{RN} \left( \sum_{j=1}^{N_0} + \sum_{N_0+1}^{R} \right) (jL_j(E)) + N^{-\frac{1}{2}+} \qquad (8.26)$$

$$< \frac{CN_0^2}{RN} + 2L(E) \frac{R}{N} + N^{-\frac{1}{2}+}$$

$$< \kappa^2 L(E) + \frac{\kappa}{5} L(E) + \kappa^{3/2} L(E) < \frac{1}{4} \kappa L(E)$$

The expression $\sum_{|j|<R} \frac{R-|j|}{R^2} u(x+j\omega) - \hat{u}(0)$ is estimated as in Theorem 5.1 and hence bounded by

$$\frac{(\log N)^3}{\kappa N^{1/2}} + \frac{1}{\kappa N} \log K + \rho(x) \tag{8.27}$$

where $\rho(x)$ satisfies

$$\int_{\mathbf{T}} |\rho(x)|^2 dx < c \sum_{|k|>K} \frac{1}{k^2} < \frac{c}{K} \tag{8.28}$$

Taking

$$\log K \sim \kappa^2 L(E) N \tag{8.29}$$

we get

$$(8.27) < K^{2-} L(E) + \frac{\kappa}{10} L(E) + \rho(x) \tag{8.30}$$

Collecting estimates (8.24), (8.25), (8.26), and (8.29), we get

$$|u(x) - L(E) \leq |L(E) - L_N(E)| + |u(x) - \hat{u}(0)|$$
$$< \tfrac{\kappa}{2} L(E) + \tfrac{\kappa}{4} L(E) + \kappa^{2-} L(E) + \tfrac{\kappa}{10} L(E) + \rho(x)$$
$$< \tfrac{9}{10} \kappa L(E) + \rho(x)$$

and (8.22) follows from (8.28) and (8.29).

Next we combine again (8.22) with the "avalanche principle," Proposition 6.1. Thus, taking $\kappa = \frac{1}{10}$ in (8.22) and $\log n \sim L(E).N$ in Proposition 6.1, we get

$$\left| \log \|M_{nN}(x, E)\| + \sum_{j=2}^{n-1} \log \|M_N(x+jN\omega, E)\| \right.$$
$$\left. - \sum_{j=1}^{n-1} \log \|M_{2N}(x+jN\omega, E)\| \right|$$
$$< ne^{-\frac{1}{2}NL(E)}$$

for $x$ outside a set of measure $< n.e^{-cL(E)N} < e^{-\frac{c}{2}L(E).N}$.

Hence, integrating

$$\left| L_{nN}(E) + \frac{n-2}{n} L_N(E) - 2\frac{n-1}{n} L_{2N}(E) \right| < e^{-cL(E)N} \tag{8.31}$$

Iterating, we thus obtain

**Lemma 8.32.** *Assume* $L(E) > 0$. *Then, for* $N$ *large enough,*

$$|L(E) + L_N(E) - 2L_{2N}(E)| < e^{-cL(E)N} \tag{8.33}$$

We emphasize here again that the constant $c > 0$ in (8.33) does not depend on $L(E)$. Clearly,

$$|\partial_E \log \|M_N(x, E)\|| \leq \sum_{j=1}^{N} \|M_{j-1}(x; E)\| \, \|M_{N-j}(x+j\omega; E)\|$$

and hence, from (8.18),

$$|\partial_E L_N(E)| \leq \max_{1 \leq j \leq N} e^{(jL_j(E)+(N-j)L_{N-j}(E)+N^{1/2+})} < e^{2NL(E)} \tag{8.34}$$

for $N$ large enough.

Assume now

$$L(E) > 0 \text{ for all } E \in [E_1, E_2]$$

Then (8.33) and (8.34) imply for $N$ sufficiently large and $E, E' \in [E_1, E_2]$

$$|L(E) - L(E')| \leq e^{-cL(E)N} + e^{-cL(E')N} + e^{2N[\max_{E'' \in [E,E']} L(E'')]}|E - E'| \tag{8.35}$$

Appropriate choice of $N$ in (8.35) shows that

$$|L(E) - L(E')| \leq |E - E'|^{c_1} \tag{8.36}$$

if $|E - E'|$ is sufficiently small, depending on (8.35), but with exponent $c_1 > 0$ independent of the smallness of the Lyapounov exponent. Together with the Thouless formula, this proves Proposition 8.3.

The preceding does not answer the question of which regularity properties $N(E)$ has without the assumption that $L(E) > 0$. The method used above permits us to show Proposition 8.4. We give a sketch of the argument.

Fix an energy $E$ s.t. $L(E) > 0$. Let $\gamma > 0$ be a small constant. Take $N_0 \in \mathbf{Z}_+$ satisfying

$$N_0^{1-\gamma} L_{N_0}(E) \geq \max_{M < N_0} (M^{1-\gamma} L_M(E), 1) \tag{8.37}$$

We may find such $N_0$ with

$$N_0 < \frac{2}{L(E)^{\frac{1}{1-\gamma}}} \tag{8.38}$$

$(L(E)$ is assumed small).

Returning to Lemma 8.18, we have in fact (see (8.19)) for $N_1 < N \leq N_0$

$$\log \|M_N(x, E)\| < NL_N(E) + N\frac{(\log N_1)^5}{N_1} + 2 \max_{\substack{j \leq N_1 \\ x \in \mathbf{T}}} \log \|M_j(x)\| \tag{8.39}$$

Take $N_1 = N_0^{-\varepsilon} N$, and iterate (8.39). This clearly gives a bound

$$\log \|M_N(x; E)\| < NL_N(E) + 2j_1 L_{j_1}(E) + \cdots + 2^{s-1} j_{s-1} L_{j_{s-1}}(E) + \\ (N_0^\varepsilon + 2N_0^\varepsilon + \cdots + 2^{s-1} N_0^\varepsilon)(\log N)^5 + C2^s \tag{8.40}$$

when

$$j_1 < N_0^{-\varepsilon} N, j_{s'} < N_0^{-\varepsilon} j_{s'-1} \text{ and hence } s < \frac{1}{\varepsilon}$$

From assumption (8.37)

$$(8.40) < N_0 L_{N_0}(E)\left(1 + 2N_0^{-\varepsilon\gamma} + \cdots + (2N_0^{-\varepsilon\gamma})^{s-1}\right)\left(\frac{N}{N_0}\right)^\gamma + 2^{\frac{1}{\varepsilon}} N_0^{\varepsilon+}$$

and taking $\varepsilon = \frac{\gamma}{2}$, again by (8.37),

$$\log \|M_N(x; E)\| < (1+\gamma)N^\gamma N_0^{1-\gamma} L_{N_0}(E) + N_0^{\frac{\gamma}{2}+} < 2N_0 L_{N_0}(E) \text{ for } N \leq N_0 \tag{8.41}$$

The proof of Lemma 8.9 also shows that

$$\left| \frac{1}{R} \sum_{j=0}^{R-1} \left( \frac{1}{N} \log \|M_N(x + j\omega; E)\| \right) - L_N(E) \right| \tag{8.42}$$

$$\leq \delta \log \frac{1}{\delta} + \frac{1}{R} (\log q_{s_*})^5 + \frac{1}{\delta q_{s_*}} + \rho_1(x)$$

with

$$\|\rho_1\|_{L^1(\mathbf{T})} \leq \|u - u_{(\delta)}\|_{L^1(\mathbf{T})} < C\delta \log \frac{1}{\delta} \tag{8.43}$$

For $\delta < \frac{1}{R}$ and $\frac{1}{\delta^2} < q_{s_*} < \frac{1}{\delta^2} (\log \frac{1}{\delta})^4$, (8.42) $\lesssim \frac{1}{R} \left( \log \frac{1}{\delta} \right)^5 + \rho_1(x)$.
Taking $N = N_0, R < N_0$, we get from the preceding

$$\left| \frac{1}{N_0} \log \|M_{N_0}(x)\| - L_{N_0}(E) \right| \lesssim \frac{1}{R} \left( \log \frac{1}{\delta} \right)^5 + \rho_1(x) + \frac{2}{N_0} \max_{j \leq R, x} \log \|M_j(x)\| \tag{8.44}$$

By (8.41), the last term of (8.44) is bounded by

$$C \left( \frac{R}{N_0} \right)^\gamma L_{N_0}(E) + N_0^{-1+\frac{\gamma}{2}+} \tag{8.45}$$

and letting $R \sim \kappa^{1/\gamma} N_0$,

$$\left| \frac{1}{N_0} \log \|M_{N_0}(x)\| - L_{N_0}(E) \right| < \kappa^{-1/\gamma} N_0^{-1} (\log \frac{1}{\delta})^5 + \kappa L_{N_0}(E) + \rho_1(x) \tag{8.46}$$

Recalling (8.43), an appropriate choice of $\delta$ then shows that

$$\left| \frac{1}{N_0} \log \|M_{N_0}(x)\| - L_{N_0}(E) \right| < o(L_{N_0}(E)) \tag{8.47}$$

for all $x$ outside a set of measure at most $e^{-N_0^{\gamma/6}}$.

One easily checks that the same statement holds with $N_0$ replaced by $\frac{1}{2}N_0$. (We assume $N_0$ even; otherwise, replace $N_0$ by $N_0 + 1$).

Recalling again (8.37), we have

$$N_0^{1-\gamma} L_{N_0}(E) > \left( \frac{N_0}{2} \right)^{1-\gamma} L_{\frac{N_0}{2}}(E)$$

and hence

$$2^{1-\gamma} L_{N_0}(E) > L_{\frac{N_0}{2}}(E) \geq L_{N_0}(E) \tag{8.48}$$

Thus, for $x$ outside a set of measure $< e^{-N_0^{\gamma/6}}$,

$$\log \|M_{\frac{N_0}{2}}(x; E)\| > \frac{N_0}{2} L_{\frac{N_0}{2}} (1 - o(1)) = \log \mu \tag{8.49}$$

and

$$\left| \log \|M_{\frac{N_0}{2}}(x)\| + \log \|M_{\frac{N_0}{2}}(x + \tfrac{N_0}{2}\omega)\| - \log \|M_{N_0}(x)\| \right| <$$
$$N_0 \left( L_{\frac{N_0}{2}}(E) - L_{N_0}(E) \right) + o(1) N_0 L_{\frac{N_0}{2}}(E) < \tag{8.50}$$
$$N_0 L_{\frac{N_0}{2}}(E) \left( 1 - 2^{\gamma-1} + o(1) \right) < \left( 1 - \tfrac{\gamma}{2} \right) \log \mu$$

Estimate (8.50) is weaker than condition (6.3) in the assumptions from Proposition 6.1. One verifies that the argument may still be carried out, provided $n < \mu^{\gamma/2}$ and with bound $n\mu^{-\gamma/2}$ in (6.4).

This thus permits us to show that

$$|L_N(E) + L_{\frac{N_0}{2}}(E) - 2L_{N_0}(E)| < c\frac{N_0}{N}L_{N_0}(E) \tag{8.51}$$

for $N < e^{\frac{1}{2}N_0^{\gamma/6}}$, $N_0|N$. In particular,

$$L_N(E) > 2L_{N_0}(E) - L_{\frac{N_0}{2}}(E) - o(\gamma)L_{N_0}(E) > \gamma L_{N_0}(E) \tag{8.52}$$

if $\gamma^{-2}N_0 < N < e^{\frac{1}{2}N_0^{\gamma/6}}$, $N_0|N$.

Continuing the argument with $N_0$ replaced by $N_1 < e^{N_0^{\gamma/10}}$ etc. shows that

$$|L(E) + L_{\frac{N_0}{2}}(E) - 2L_{N_0}(E)| < e^{-N_0^{\gamma/10}} \tag{8.53}$$
$$L(E) > \gamma L_{N_0}(E) \tag{8.54}$$

Also, if $N_0|N$ and $L_N(E) - L_{2N}(E) < \frac{1}{10}L_N(E)$, we have

$$|L(E) + L_N(E) - 2L_{2N}(E)| < e^{-N^{\gamma/10}} \tag{8.55}$$

Recalling (8.41), it follows from subadditivity, (8.38) and (8.54), that

$$\max_x(\log \|M_N(x, E)\|) < 4\gamma^{-1}NL(E) \text{ if } N > L(E)^{-\frac{1}{1-\gamma}}$$
$$\max_x(\log \|M_N(x, E)\|) < 4\gamma^{-1}(N + L(E)^{-\frac{1}{1-\gamma}})L(E) \text{ for all } N \tag{8.56}$$

Next, take $E'$ such that

$$\tau \equiv |E - E'| < e^{-C(\gamma)L(E)^{1-\frac{1}{1-\gamma}}} \tag{8.57}$$

with $C(\gamma)$ sufficiently large. One may then find $N > \dfrac{1}{L(E)^{\frac{1}{1-\gamma}}}$ satisfying the conditions

$$N_0 = N_0(E)|N$$

$$e^{-C(\gamma)NL(E)} < |E - E'| < e^{-\frac{N}{\gamma^2}}L(E) \tag{8.58}$$

$$L_N(E) - L_{2N}(E) < \frac{1}{10}L_N(E) \tag{8.59}$$

Estimate, using (8.56), for $M \leq 2N$

$$|L_M(E) - L_M(E')| \leq |E - E'| \left[ \max_{\substack{j \leq M, x \\ |E - E''| < \tau}} \|M_j(x, E'')\|^2 \right]$$
$$< |E - E'| \exp\left[ 8\gamma^{-1} \max_{|E - E''| < \tau} (2N + L(E'')^{-\frac{1}{1-\gamma}})L(E'') \right] \tag{8.60}$$

Assume that

$$L(E) \sim L(E'') \text{ for all } |E - E''| \leq |E - E'| \tag{8.61}$$

It follows then from (8.58) and (8.60) that

$$|L_M(E) - L_M(E')| < |E - E'|^{1/2} < o(L_M(E)) \text{ for } M \leq 2N \tag{8.62}$$

Since from (8.62) we may in (8.37) take $N_0(E') = N_0(E)$, inequality (8.55) holds for both $E$ and $E'$. Therefore, collecting estimates,

$$|L(E) - L(E')| < 2e^{-N^{\gamma/10}} + |L_N(E) - L_N(E')| + 2|L_{2N}(E) - L_{2N}(E')|$$
$$< 2e^{-N^{\gamma/10}} + 3|E - E'|^{1/2}$$
$$< 3e^{-N^{\gamma/10}}$$
$$< e^{-(\log \frac{1}{|E-E'|})^{\gamma/10}} \tag{8.63}$$

Recalling (8.57) and the fact that $L(E)$ is small, it follows that $(8.63) = o(1)L(E)$.

Thus, if (8.57) and (8.61) hold, then $|L(E) - L(E'')| = o(L(E))$ and (8.63) hold. Hence (8.57) implies (8.63).

Given energies $E, E'$ s.t. $L(E) \geq L(E'), L(E) > 0$, either (8.57) and (8.63) hold, or if (8.57) fails,

$$|L(E) - L(E')| \leq L(E) < C(\gamma) \left( \log \frac{1}{|E - E'|} \right)^{-\frac{1}{2\gamma}} \tag{8.64}$$

Hence (8.64) is valid in either case.

Since $\gamma > 0$ was arbitrarily small, it follows that for any $A > 0$

$$|L(E) - L(E')| \leq C_A \left( \log \frac{1}{|E - E'|} \right)^{-A}$$

and Proposition 8.4 follows again from the Thouless formula.

**Remark.** Assume the energy satisfies

$$\overline{\lim} \frac{\log N L_N(E)}{\log N} > 0 \tag{8.65}$$

We may then find (arbitrarily large) $N_0 \in \mathbf{Z}_+$ satisfying (8.37) for some $\gamma > 0$. From (8.54), $L(E) > 0$. Thus (8.65) implies positivity of the Lyapounov exponent.

# *References*

[C-S]  W. Craig, B. Simon.  Log Hölder continuity of the integrated density of states for stochastic Jacobi matrices, *CMP* 90(2) (1983), 207–218.

[H-S]  B. Helffer, J. Sjostrand.  Analyse semi-classique pour l'equation de Harper, *Memoire SMF* 34 (1988).

# Chapter Nine

---

## Some Facts about Semialgebraic Sets

The purpose of this chapter is to summarize a number of results from the literature for later use. Some slightly weaker statements (which do suffice for our needs) also may be found in Bourgain and Goldstein [5] (in Related References) with proof.

**Definition 9.1.** A set $S \subset \mathbf{R}^n$ is called *semialgebraic* if it is a finite union of sets defined by a finite number of polynomial equalities and inequalities. More precisely, let $\mathcal{P} = \{P_1, \ldots, P_s\} \subset \mathbf{R}[X_1, \ldots, X_n]$ be a family of real polynomials whose degrees are bounded by $d$. A (closed) semialgebraic set $S$ is given by an expression

$$S = \bigcup_j \bigcap_{\ell \in \mathcal{L}_j} \{\mathbf{R}^n | P_\ell s_{j\ell} 0\},$$

where $\mathcal{L}_j \subset \{1, \ldots, s\}$ and $s_{j\ell} \in \{\geq, \leq, =\}$ are arbitrary. We say that $S$ has degree at most $sd$, and its degree is the infimum of $sd$ over all representations as in (9.1).

The projection of a semialgebraic set of $\mathbf{R}^{k+\ell}$ onto $\mathbf{R}^k$ is semialgebraic. This is known as the *Tarski-Seidenberg principle*; see Bochnak, Coste, and Roy [4]. The currently best quantitative version of this principle is due to Basu, Pollack, and Roy [3] and Basu [1]. For the history of such effective Tarski-Seidenberg results, we refer the reader to those papers.

**Proposition 9.2.** Let $S \subset \mathbf{R}^n$ be semialgebraic defined in terms of $s$ polynomials of degree at most $d$ as in (9.1). Then there exists a semialgebraic description of its projection onto $\mathbf{R}^{n-1}$ by a formula involving at most $s^{2n} d^{O(n)}$ polynomials of degree at most $d^{O(n)}$. In particular, if $S$ has degree $B$, then any projection of $S$ has degree at most $B^C, C = C(n)$.

This is a special case of the main theorem in Basu, Pollack, and Roy [3].

Another fundamental result on semialgebraic sets is the following bound on the sum of the Betti numbers by Milnor, Oleinik, and Petrovsky, and Thom. Strictly speaking, their result applies only to *basic* semialgebraic sets, which are given purely by intersections without unions. The general case as in Definition 9.1 above was settled by Basu [2].

**Theorem 9.3.** Let $S \subset \mathbf{R}^n$ be as in (9.1). Then the sum of all Betti numbers of $S$ is bounded by $s^n (O(d))^n$. In particular, the number of connected components of $S$ does not exceed $s^n (O(d))^n$.

This is a special case of Theorem 1 in Basu [2].

Another result that we shall need is the following triangulation theorem of Yomdin [7], later refined by Gromov [6]. We basically reproduce the statement of that result from Gromov [6; see p. 239].

**Theorem 9.4.** For any positive integers $r, n$ there exists a constant $C = C(n, r)$ with the following property: Any semialgebraic set $S \subset [0, 1]^n \subset \mathbf{R}^n$ can be triangulated into $N \lesssim (\deg \mathcal{S} + 1)^C$ simplices, where for every closed $k$-simplex $\Delta \subset \mathcal{S}$ there exists a homeomorphism $h_\Delta$ of the regular simplex $\Delta^k \subset \mathbf{R}^k$ with unit edge length onto $\Delta$ such that $h_\Delta$ is real analytic in the interior of each face of $\Delta$. Furthermore, $\|D_r h_\Delta\| \leq 1$ for all $\Delta$.

### Related References

[1] S. Basu. New results on quantifier elimination over real closed fields and applications to constraint databases, *J. ACM* 46(4) (1999), 537–555.

[2] S. Basu. On bounding the Betti numbers and computing the Euler characteristic of semi-algebraic sets, *Discrete Comput. Geom.* 22(1) (1999), 1–18.

[3] S. Basu, R. Pollack, M.-F. Roy. On the combinatorial and algebraic complexity of quantifier elimination, *J. ACM* 43(6) (1996), 1002–1045.

[4] J. Bochnak, M. Coste, M.-F. Roy. Real algebraic geometry, *Ergebnisse der Mathematik und ihrer Grenzgebiete* (3), 36, Springer-Verlag, Berlin, 1998.

[5] J. Bourgain, M. Goldstein. On non-perturbative localization with quasi-periodic potential, *Annals of Math.* (2) 152(3) (2000), 835–879.

[6] M. Gromov. Entropy, homology and semialgebraic geometry, *Séminaire Bourbaki*, 1985/86(663), *Astérisque* 145–146 (1987), 5, 225–240.

[7] Y. Yomdin. $C^k$-resolution of semi-algebraic mappings. *Israel J. Math.*, 57(3) (1987), 301–317.

In our applications, the number $n$ of variables always will be bounded. Returning to Theorem 9.4, we notice the following corollaries.

**Corollary 9.5.**

(i)  Let $\mathcal{S} \subset [0,1]^n$ be connected and semialgebraic of degree $B$. If $p, q \in \mathcal{S}$, there is a path $\gamma : [0,1] \to \mathcal{S}$ s.t. $\gamma(0) = p, \gamma(1) = q$ and $|\dot{\gamma}| < B^C$.

(ii) Let $\mathcal{S} \subset \prod_{j=1}^n [0, \rho_j]$. Then, for $p, q \in \mathcal{S}$, there is a path $\gamma : [0,1] \to \mathcal{S}$ s.t. $\gamma(0) = p, \gamma(1) = q$ and

$$\sum_{j=1}^n \frac{|\dot{\gamma}_j|}{\rho_j} < B^C$$

((ii) follows from simple rescaling).

**Corollary 9.6.** Let $\mathcal{S} \subset [0,1]^n$ be semialgebraic of degree $B$. Let $\varepsilon > 0$ be a small number and $\mathrm{mes}_n \mathcal{S} < \varepsilon^n$. Then $\mathcal{S}$ may be covered by at most $B^C (\frac{1}{\varepsilon})^{n-1}$ $\varepsilon$-balls.

**Proof.** From the assumption, clearly, $\mathrm{dist}(p, \partial \mathcal{S}) < \varepsilon$ for all $p \in \mathcal{S}$. By Theorem 9.4, $\partial \mathcal{S}$ is obtained as an image of at most $B^C$ sufficiently smooth maps defined on regular simplices $\Delta^k$ with $k \leq n - 1$. The conclusion is clear.

The next fact deals with the intersection of a semialgebraic set of small measure and the orbit of a diophantine shift.

**Corollary 9.7.** *Let $S \subset [0,1]^n$ be semialgebraic of degree $B$ and $\mathrm{mes}_n S < \eta$. Let $\omega \in \mathbf{T}^n$ satisfy a DC and $N$ be a large integer,*

$$\log B \ll \log N < \log \frac{1}{\eta}$$

*Then, for any $x_0 \in \mathbf{T}^n$*

$$\#\{k = 1, \cdots, N | x_0 + k\omega \in S(\mathrm{mod}\, 1)\} < N^{1-\delta} \tag{9.8}$$

*for some $\delta = \delta(\omega)$.*

**Proof.** Choose $\varepsilon > \eta^{1/n} + N^{-\delta}$. Then $S$ is covered by at most $B^C(\frac{1}{\varepsilon})^{n-1}$ $\varepsilon$-balls, and the orbit occupation of a single ball is at most $C\varepsilon^n N$ (since $N$ is large enough). This follows from the fact that $\omega$ was assumed diophantine and standard equidistribution considerations.

The statement may be seen as follows. Let $\chi$ be the indicator function of the ball $B(0, \varepsilon)$ centered at 0. If $R = \frac{1}{10\varepsilon}$ and $F_R$ is the usual (1-dim) Fejer-kernel, one has that $\chi \leq CR^{-n} \prod_{j=1}^{n} F_R(x_j)$. Therefore, passing to Fourier transform,

$$\sum_1^N \chi(x_o + k\omega) \leq CR^{-n} \sum_{0 \leq |\ell_1|, \dots, |\ell_n| < R} \widehat{F}_R(\ell_1) \cdots \widehat{F}_R(\ell_n) \left| \sum_1^N e^{ik\ell.\omega} \right|$$

$$\leq CR^{-n}N + C \max_{0 < |\ell| \lesssim R} |1 - e^{i\ell.\omega}|^{-1}$$

$$< C\varepsilon^n N + CR^A < CN\left(\varepsilon^n + \frac{\varepsilon^{-A}}{N}\right)$$

(where the constant $A$ depends on $\omega$).

Thus we get the bound

$$B^C \left(\frac{1}{\varepsilon}\right)^{n-1} \varepsilon^n N < B^C \varepsilon N < N^{1-\delta}$$

Finally, we also will use the following fact:

**Lemma 9.9.** *Let $S \subset [0,1]^{2n}$ be a semialgebraic set of degree $B$ and $\mathrm{mes}_{2n} S < \eta$, $\log B \ll \log \frac{1}{\eta}$. We denote $(\omega, x) \in [0,1]^n \times [0,1]^n$ the product variable. Fix $\varepsilon > \eta^{\frac{1}{2n}}$. Then there is a decomposition*

$$S = S_1 \cup S_2$$

*$S_1$ satisfying*

$$\mathrm{Proj}_\omega S_1 < B^C \varepsilon \tag{9.10}$$

*and $S_2$ satisfying the transversality property*

$$\mathrm{mes}_n(S_2 \cap L) < B^C \varepsilon^{-1} \eta^{1/2n} \tag{9.11}$$

*for any $n$-dimensional hyperplane $L$ s.t. $\max_{0 \leq j \leq n-1} |\mathrm{Proj}_L(e_j)| < \frac{1}{100}\varepsilon$ (we denote $(e_0, \dots, e_{n-1})$ the $\omega$-coordinate vectors).*

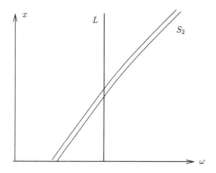

**Proof.** We treat the case $n = 1$ for simplicity, but the argument generalizes; see [BGS]. Since $\operatorname{dist}(p, \partial S) < \eta_1 = \eta^{\frac{1}{2n}}$ for all $p \in S$, we may replace $S$ by a $\eta_1$-neighborhood $\Gamma_{\eta_1}$ of the zero set $\Gamma = [P(\omega, x) = 0]$, $P$ a polynomial of degree at most $B$.

Define

$$\Gamma' = \Gamma \cap [|\partial_x P| < \varepsilon |\partial_\omega P|]$$
$$\Gamma'' = \Gamma \cap [|\partial_x P| \geq \varepsilon |\partial_\omega P|]$$
$$S_1 = \Gamma'_{\eta_1}; S_2 = \Gamma''_{\eta_1}$$

Thus $\operatorname{Proj}_\omega \Gamma'$ is a union of at most $B^C$-intervals obtained as the $\gamma_0$-range, $\gamma = (\gamma_0, \gamma_1) : [0,1] \to \Gamma'$ a curve satisfying $|\dot{\gamma}| \leq B^C$. Since $P(\gamma_0(s), \gamma_1(s)) = 1$, it follows from the definition of $\Gamma'$ that

$$(\partial_\omega P)(\gamma(s))\dot{\gamma}_0(s) + (\partial_x P)(\gamma(s))\dot{\gamma}_1(s) = 0$$
$$|\dot{\gamma}_0(s)| \leq \left(\max_s \frac{|\partial_x P|}{|\partial_\omega P|}\right) B^C < \varepsilon B^C$$
$$\text{range } \gamma_0 = \int_0^1 |\dot{\gamma}_0(s)| < \varepsilon B^C$$

This gives (9.10).

Next, we prove (9.11). Thus $L$ is a segment satisfying

$$|\vec{L}.e_0| < \frac{\varepsilon}{100} \tag{9.12}$$

The set $S_2 \cap L = \Gamma''_{\eta_1} \cap L$ is contained in an $\eta_1$-neighborhood of $\operatorname{Proj}_L(\Gamma'' \cap L_{\eta_1})$. The set $\Gamma'' \cap L_{\eta_1}$ has at most $B^C$ connected components.

From Corollary 9.5 (ii) and (9.12), a pair of points in one component may be joined by a curve $\gamma = (\gamma_0, \gamma_1) : [0,1] \to \Gamma'' \cap L_{\eta_1}$ where

$$\begin{pmatrix} \gamma_0 \\ \gamma_1 \end{pmatrix} = \begin{pmatrix} \cos\theta & -\sin\theta \\ \sin\theta & \cos\theta \end{pmatrix} \begin{pmatrix} \beta_0 \\ \beta_1 \end{pmatrix}$$
$$\frac{|\dot{\beta}_0|}{\eta_1} + |\dot{\beta}_1| < B^C, |\theta| < \frac{\varepsilon}{100} \text{ depending on } L$$

Write again

$$P(\gamma(s)) = P(\cos\theta.\beta_0 - \sin\theta.\beta_1, \sin\theta.\beta_0 + \cos\theta.\beta_1) = 0$$

and differentiating in $S$, we get

$$\left\langle \begin{pmatrix} \cos\theta & -\sin\theta \\ \sin\theta & \cos\theta \end{pmatrix} \begin{pmatrix} \dot{\beta}_0 \\ \dot{\beta}_1 \end{pmatrix}, \begin{pmatrix} \partial_\omega P \\ \partial_x P \end{pmatrix} \right\rangle = 0$$

Hence

$$|\dot{\beta}_1| < \frac{\cos\theta |\partial_\omega P| + |\sin\theta| \, |\partial_x P|}{|-\sin\theta . \partial_\omega P + \cos\theta . \partial_x P|} |\dot{\beta}_0| < \frac{|\partial_\omega P| + |\theta| \, |\partial_x P|}{(1-\theta^2)|\partial_x P| - |\theta| \, |\partial_\omega P|} |\dot{\beta}_0|$$

$$< \frac{\frac{1}{\varepsilon} + |\theta|}{1 - \theta^2 - \frac{|\theta|}{\varepsilon}} B^C \eta_1 < \frac{2}{\varepsilon} B^C \eta_1$$

by definition of $\Gamma''$. Therefore,

$$\text{mes Proj}_L (\Gamma'' \cap L_{\eta_1}) < \frac{1}{\varepsilon} B^C \eta_1$$

and

$$\text{mes}\,(\mathcal{S}_2 \cap L) < \frac{1}{\varepsilon} B^C \eta^{1/2n}$$

This proves Lemma 9.9.

## *Reference*

[BGS] J. Bourgain, M. Goldstein, W. Schlag. Anderson localization for Schrödinger operators on $\mathbf{Z}^2$ with quasi-periodic potential, *Acta Math.* 188 (2002), 41–86.

# Chapter Ten

## Localization

We will prove the following theorem.

**Theorem 10.1.** *Consider the 1D lattice Schrödinger operator*

$$H_\omega(x) = v(x + n\omega)\delta_{nn'} + \Delta$$

*where $v$ is an analytic potential on $\mathbf{T}^d$ and $\omega \in DC = DC_{A,c} \subset \mathbf{T}^d$ refers to frequency vectors satisfying a diophantine condition*

$$\|k.\omega\| > c|k|^{-A} \text{ for } k \in \mathbf{Z}^d\backslash\{0\} \tag{10.2}$$

*Assume that the Lyapounov exponent*

$$L(E) = L_\omega(E) > c_0 \tag{10.3}$$

*for all $\omega \in DC$ and $E \in \mathbf{R}$.*

*Fix $x_0 \in \mathbf{T}^d$. Then, for almost all $\omega \in DC$, $H_\omega(x_0)$ satisfies Anderson localization (i.e., $H$ has p.p. spectrum with exponentially localized states).*

**Remarks.**
1. The case $d = 1, 2$ was treated in [B-G] (see Chapter 1). The argument presented below makes more extensive use of semialgebraic set theory to deal with certain transversality issues and applies to arbitrary $d$.
2. This result is *nonperturbative*. Assume, for instance, that $v_0$ is a nonconstant trigonometric polynomial on $\mathbf{T}^d$, and let $v = \lambda v_0$. Herman's lower bound permits us then to specify $\lambda \geq \lambda_0(v)$ for (10.3) to hold, independently of $\omega$.
3. In Theorem 10.1, the frequency vector $\omega$ also is considered as a parameter. The set of "good $\omega$'s" in the conclusion of Theorem 10.1 requires, at least in our argument, a further exclusion of a zero-measure set, for which we do not have a simple arithmetic description. In some cases, for instance, in the Almost Mathieu case, $v(x) = \lambda \cos x$, an explicit arithmetic condition may be stated. We will discuss this at the end of this chapter.

There are the following two main ingredients in the proof of Theorem 10.1:
1. The LDT for the fundamental matrix and Proposition 7.19 on the Green's function estimate
2. Semialgebraic set theory as described in Chapter 9

As was already mentioned, to establish AL for $H$, it suffices to show that any extended state is exponentially decaying. Thus, if $\xi = (\xi_n)_{n\in\mathbf{Z}}$ and $E \in \mathbf{R}$ satisfy

$$|\xi_n| < C|n| \text{ for } |n| \to \infty$$

and

$$H\xi = E\xi$$

then

$$|\xi_n| < e^{-c|n|} \text{ for } |n| \to \infty \tag{10.4}$$

(In fact, it can be shown that the exponent $c$ in (10.4) may be taken $L(E) - \varepsilon$ for all $\varepsilon > 0$.)

Fix $N_0$ and consider the property

$$|G_{N_0}(E, x)(n_1, n_2)| < e^{-c_0|n_1 - n_2| + N_0^{1-}} \text{ for all } 1 \le n_1, n_2 \le N_0 \tag{10.5}$$

Writing $v(x) = \sum_{k \in \mathbb{Z}^d} \hat{v}(k) e^{ik.x}$, $|\hat{v}(k)| < e^{-\rho|k|}$, it is clear that in (10.5) we basically may substitute $v$ by $v_1(x) = \sum_{|k| < CN_0} \hat{v}(k) e^{ik.x}$. Replace (10.5) by the condition

$$\sum_{1 \le n_1, n_2 \le N_0} e^{2c_0|n_1 - n_2|} [\det(H_{N_0} - E)_{n_1, n_2}]^2 \le e^{2N_0^{1-}} [\det(H_{N_0} - E)]^2 \tag{10.6}$$

(where $A_{n_1, n_2}$ denotes the $(n_1, n_2)$-minor of the matrix $A$). Clearly, (10.6) is of the form

$$P(\cos \omega, \sin \omega, \cos x, \sin x, E) \ge 0 \tag{10.7}$$

where $P$ is a polynomial of degree at most $CN_0^2$.

One may further (assuming $x$ bounded) truncate power series for "cos" and "sin" and replace (10.7) by a polynomial

$$P(\omega, x, E) \ge 0 \tag{10.8}$$

of degree at most say $N_0^3$.

Therefore, fixing $\omega \in DC$ and $E$ and returning to the statement in Proposition 7.19, the exponential set $\Omega$ does not only satisfy the measure estimate

$$\text{mes } \Omega < e^{-cN_0^\sigma} \tag{10.9}$$

but also may be assumed semialgebraic of degree $< N_0^3$. Notice that this set depends on $E$ (and on $\omega$).

The main issue in what follows is to eliminate the energy $E$. We explain the idea. Fix $N = N_0$ in Proposition 7.19 and redefine $\Lambda$ to be one of the intervals

$$\Lambda = [-N_0, N_0], [-N_0, N_0 - 1], [-N_0 + 1, N_0], [-N_0 + 1, N_0 - 1] \tag{10.10}$$

Let $\Omega = \Omega(E)$ be as above. Thus, if $x \notin \Omega$, then one of the intervals $\Lambda$ (depending on $x$) satisfies

$$|G_\Lambda(E, x)(n_1, n_2)| < e^{-c_0|n_1 - n_2| + N_0^{-1}} \text{ for all } n_1, n_2 \in \Lambda \tag{10.11}$$

Fix $x_0 \in \mathbb{T}^d$ and consider the orbit $\{x_0 + j\omega \mid |j| \le N_1\}$ where

$$N_1 = N_0^C$$

($C$ a sufficiently large constant).

Apply Corollary 9.7 with $\mathcal{S} = \Omega$, $B = N_0^3$, $\eta = e^{-cN_0^\sigma}$ and $N = N_1$. The conclusion (9.8) implies that except for at most $N_1^{1-\delta}$ values of $|j| < N_1$, taking $x = x_0 + j\omega$, one of the intervals $\Lambda$ from (10.10) satisfies (10.11).

Recall that

$$(H(x_0) - E)\xi = 0$$

and therefore, if $\Lambda + j = [a, b]$

$$(R_{\Lambda+j}(H - E)R_{\Lambda+j})\xi = -(R_{\Lambda+j}HR_{Z\setminus(\Lambda+j)})\xi = -(\xi_{a-1}e_a + \xi_{b+1}e_b)$$

Hence, for $n \in \Lambda + j$,

$$|\xi_n| < \max_{m \in \{a,b\}} |G_{\Lambda+j}(x_0, E)(m, n)| \cdot \Big( \max_{|k| < N_1 + N_0} |\xi_k| \Big)$$

$$< \max_{m \in \{a,b\}} |G_\Lambda(x_0 + j\omega, E)(m - j, n - j)|N_1$$

$$< N_1 . e^{N_0^{1-}} (e^{-c_0|a-n|} + e^{-c_0|b-n|})$$

Taking in particular $n = j$, we have $|j - a| > \frac{N_0}{2}, |j - b| > \frac{N_0}{2}$ by (10.10), and thus

$$|\xi_j| < e^{-\frac{c_0}{2} N_0} \tag{10.12}$$

holds for all $|j| < N_1$ except $N_1^{1-\delta}$ many.

Next, let

$$I = [-j_0 + 1, j_0 - 1]$$

and write

$$R_I (H(x_0) - E)R_I\xi = -(\xi_{-j_0}e_{-j_0+1} + \xi_{j_0}e_{j_0-1})$$
$$1 = |\xi_0| \leq |G_I(x_0, E)(0, j_0 - 1)| \, |\xi_{j_0}| + |G_I(x_0, E)(0, -j_0 + 1)| \, |\xi_{-j_0}|$$
$$\leq \|G_I(x_0, E)\|(|\xi_{j_0}| + |\xi_{-j_0}|)$$

If $j_0, -j_0$ satisfy both (10.12), we conclude that

$$\|G_{]-j_0,j_0[}(x_0, E)\| > e^{\frac{c_0}{2} N_0} \tag{10.13}$$

or equivalently

$$\text{dist}\big(E, \, \text{Spec} H_{]-j_0,j_0[}(x_0)\big) < e^{-\frac{c_0}{2} N_0} \tag{10.14}$$

Thus, if there is an extended state $\xi, \xi_0 = 1$ with energy $E$, then, for any large $N_0$, there is some $j_0, |j_0| < N_1 = N_0^C$ for which (10.14) holds.

Denote

$$\mathcal{E} = \mathcal{E}_\omega = \bigcup_{|j| \leq N_1} \text{Spec} \, H_{]-j,j[}(x_0) \tag{10.15}$$

It clearly follows from (10.11) and (10.14) that if

$$x \notin \bigcup_{E' \in \mathcal{E}_\omega} \Omega(E') \tag{10.16}$$

then one of the sets $\Lambda$ in (10.10) satisfies

$$|G_\Lambda(E, x)(n_1, n_2)| < e^{-c_0|n_1 - n_2| + N_0^{1-}} \text{ for } n_1, n_2 \in \Lambda \tag{10.17}$$

Next, let

$$N_2 = N_0^{C'}$$

with $C'$ a sufficiently large constant, and suppose that we ensured that

$$x_0 + n\omega \notin \bigcup_{E' \in \mathcal{E}_\omega} \Omega(E') \pmod 1 \text{ for all } N_2^{1/2} < |n| < N_2 \qquad (10.18)$$

Thus, for each $N_2^{1/2} < |n| < 2N_2$, there is an interval

$$\Lambda^{(n)} \in \{[-N_0, N_0], [-N_0, N_0 - 1], [-N_0 + 1, N_0], [-N_0 + 1, N_0 - 1]\}$$

for which (10.17) holds:

$$|G_{\Lambda^{(n)}+n}(E, x_0)(n_1, n_2)| < e^{-c_0|n_1 - n_2| + N_0^{1-}} \text{ for } n_1, n_2 \in \Lambda^{(n)} + n \qquad (10.19)$$

Define the interval

$$\widetilde{\Lambda} = \bigcup_{N_2^{1/2} < n < 2N_2} (\Lambda^{(n)} + n) \supset [N_2^{1/2}, 2N_2]$$

Application of the resolvent identity (details will be given later) permits us then to deduce from (10.19) that

$$|G_{\widetilde{\Lambda}}(E, x_0)(n_1, n_2)| < e^{-(c_0-)|n_1 - n_2|} \text{ if } |n_1 - n_2| > \frac{1}{10}N_2 \qquad (10.20)$$

and therefore,

$$|\xi_j| < e^{-(c_0-)j} \text{ if } \frac{1}{2}N_2 \le j \le N_2$$

(similarly for $j \in \mathbf{Z}_-$). Consequently, we obtain the exponential decay (10.1).

Returning to condition (10.18), consider for $|j| \le N_1$ the set

$$\mathfrak{S} = \mathfrak{S}_j \subset \mathbf{T}^d \times \mathbf{R} \times \mathbf{T}^d$$

of triplets $(\omega, E', x)$ where

$$\omega \in DC \qquad (10.21)$$

$$E' \in \operatorname{Spec} H_{]-j,j[}(x_0) \qquad (10.22)$$

$$x \in \Omega(E') \qquad (10.23)$$

Observe that if we fix $\omega$, then the set (10.22) is specified and hence the $2j - 1$ values of $E'$. For each of these, the set $\Omega(E')$ (also dependent on $\omega$) is of measure at most $e^{-cN_0^\sigma}$. Thus we obtain that

$$S = \operatorname{Proj}_{(\omega, x)} \mathfrak{S}_j \subset \mathbf{T}^d \times \mathbf{T}^d \qquad (10.24)$$

satisfies

$$\operatorname{mes} S < e^{-cN_0^\sigma} \qquad (10.25)$$

Observe that the role of (10.21) is to ensure the LDT at scale $N_0$, and in fact, it suffices to assume (10.2) for $|k| < N_0^2$ (see Chapter 5). Therefore, we may replace in (10.21) the set $\Omega$ by a union of intervals in $\mathbf{T}^d$ of size $N_0^{-A-2}$ say.

Consider condition (10.22); thus

$$\det[R_{]-j,j[}(H_\omega(x_0) - E')R_{]-j,j[}] = 0 \qquad (10.26)$$

As before, having replaced $v$ by a trigonometric polynomial $v_1$ of degree $< CN_0$ (which is permitted in (10.13)), (10.26) becomes a polynomial condition

$$P_1(\omega, E') = 0 \tag{10.27}$$

of degree at most $j^3 < N_1^3$. It was already pointed out that (10.23) may be replaced by a condition

$$P_2(\omega, \{x\}, E') \geq 0 \tag{10.28}$$

with $P_2$ of degree at most $N_0^3$ (where $\{x\} \in [0,1]^d$ refers to $x(\mathrm{mod}\,1)$).

In conclusion, $\mathfrak{S}$ is a semialgebraic set of degree at most $N_1^3$, and applying Proposition 9.2, its projection $S$ is semialgebraic of degree at most $N_1^{3C_0}$ (for some constant $C_0$).

Returning to condition (10.18), we need to impose that

$$(\omega, \{x_0 + n\omega\}) \notin S \tag{10.29}$$

for $N_2^{1/2} \leq |n| \leq 2N_2$. To ensure (10.29), a further restriction in $\omega$ will be made.

We apply Lemma 9.9. Thus $n = d$, $B = N_1^{3C_0}$, $\eta = e^{-cN_0^\sigma}$, by (10.25).

Take $\varepsilon = N_2^{-\frac{1}{10}}$ and perform the decomposition

$$S = S_1 \cup S_2$$

from Lemma 9.9. Thus

$$\mathrm{Proj}_\omega S_1 < B^C \varepsilon < N_1^{3C_0 C} N_2^{-\frac{1}{10}} < N_2^{-\frac{1}{11}} \tag{10.30}$$

if $N_2$ is chosen large enough.

Partition

$$[0,1]^d = \bigcup_\alpha \left( x_\alpha + \left[0, \frac{1}{N_2}\right]^d \right) \qquad \alpha \leq N_2^d$$

For fixed $N_2^{1/2} \leq |n| \leq 2N_2$ and $\alpha$, consider

$$L = \left[ (\omega, x_0 + \{n x_\alpha\} + n\omega) \big| \omega \in \left[0, \frac{1}{N_2}\right]^d \right]$$

which is a translate of the $d$-hyperplane

$$\left[ \frac{1}{n} e_j + e_{j+d} \big| 0 \leq j \leq d-1 \right]$$

that satisfies the almost-orthogonality condition, since $|n| > N_2^{1/2}$,

$$\max_{0 \leq j < d} |\mathrm{Proj}_L e_j| < \varepsilon^2$$

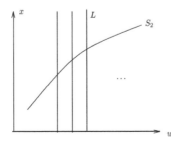

From the transversality property (9.11), it follows that

$$\operatorname{mes}_{\mathbf{T}^d}\left\{\omega \in \left[0, \tfrac{1}{N_2}\right]^d \,\middle|\, (\omega, x_0 + \{n x_\alpha\} + n\omega) \in S_2\right\} = \operatorname{mes}\,(S_2 \cap L)$$
$$< B^C \varepsilon^{-1} \eta^{1/2d}$$
$$< N_1^{3C_0 C} N_2^{1/10} e^{-\frac{c}{2d} N_0^\sigma}$$

and summing the contributions over $n$ and $\alpha$

$$\operatorname{mes}\{\omega \,|\, (\omega, \{x_0 + n\omega\}) \in S_2 \text{ for some } N_2^{1/2} < |n| < 2N_2\} <$$
$$C N_2^{d+1} N_1^{3C_0 C} N_2^{1/10} e^{-\frac{c}{2d} N_0^\sigma} < e^{-N_0^{\sigma/2}} \tag{10.31}$$

Hence (10.30) and (10.31) exclude an $\omega$-set of measure at most $N_2^{-\frac{1}{11}} + e^{-N_0^{\sigma/2}} < N_2^{-1/12}$. It remains to sum these exceptional sets over the different $j$-values, $|j| \le N_1$, to finally get the removal of an $\omega$-set of measure $< N_1 N_2^{-1/12} < N_2^{-1/13}$.

Recall that initially $N_0$ was fixed and arbitrary, and $N_1 = N_0^C$, $N_2 = N_0^{C'}$ with appropriate constants $C, C'$. Excluding thus an $\omega$-set $\mathcal{R}_{(N_0)}$ of measure

$$\operatorname{mes}\mathcal{R}_{(N_0)} < N_0^{-10}$$

we ensure that whenever

$$H_\omega(x_0)\xi = E\xi \quad |\xi_n| \lesssim |n| \text{ and } \xi_0 = 1$$

then

$$|\xi_j| < e^{-(c_0-)|j|} \text{ for } \tfrac{1}{2} N_2 \le |j| \le N_2$$

Hence, if

$$\mathcal{R} = \bigcup_N \bigcap_{N_0 > N} \mathcal{R}_{(N_0)}$$

$\operatorname{mes}\mathcal{R} = 0$ and for $\omega \in DC \backslash \mathcal{R}$, Anderson localization holds. This proves Theorem 10.1.

It remains to specify the use of the resolvent identity in establishing (10.20). Recall that if

$$\Lambda = \Lambda_1 \cup \Lambda_2, \ \Lambda_1 \cap \Lambda_2 = \phi$$

are finite subsets of $\mathbf{Z}$, then

$$G_\Lambda = (G_{\Lambda_1} + G_{\Lambda_2}) - (G_{\Lambda_1} + G_{\Lambda_2})(H_\Lambda - H_{\Lambda_1} - H_{\Lambda_2})G_\Lambda$$

(provided the relevant matrices $R_\Lambda(H-E)R_\Lambda$ and $R_{\Lambda_i}(H-E)R_{\Lambda_i}$ are invertible).

In our application, $\Lambda_1$ will be a subinterval of a larger interval $\Lambda$. If $m \in \Lambda_1, n \in \Lambda$, we have

$$|G_\Lambda(m,n)| \le |G_{\Lambda_1}(m,n)|\chi_{\Lambda_1}(n) + \sum_{\substack{n' \in \Lambda_1 \\ n'' \in \Lambda_2, |n'-n''|=1}} |G_{\Lambda_1}(m,n')|\,|G_\Lambda(n'',n)| \tag{10.32}$$

Deriving (10.20) from (10.21) is immediate from the following.

**Lemma 10.33.** Let $I \subset \mathbf{Z}$ be an interval of size $N$ and $\{I_\alpha\}$ subintervals of size $M \ll N$. Assume that

(i) *If $k \in I$, then there is some $\alpha$ s.t.*

$$\left[ k - \frac{M}{4}, k + \frac{M}{4} \right] \cap I \subset I_\alpha$$

(ii) *For all $\alpha$*

$$\|G_{I_\alpha}\| < e^{M^{1-}}$$

*and*

$$|G_{I_\alpha}(n_1, n_2)| < e^{-c_0|n_1 - n_2|} \text{ for } n_1, n_2 \in I_\alpha, |n_1 - n_2| > \frac{M}{10}$$

*Then*

$$|G_I(n_1, n_2)| < e^M \text{ and } |G_I(n_1, n_2)| < e^{-(c_0-)|n_1 - n_2|} \text{ if } n_1, n_2 \in I, |n_1 - n_2| > \frac{N}{10}$$

**Proof.** Iterate (10.32).

Putting Theorem 10.1 together with Herman's method and Proposition 7.2 on Lyapounov minorations, we get

**Corollary 10.34.** *Consider $H_\omega(x) = \lambda v_0(x + n\omega)\delta_{nn'} + \Delta$, where $v_0$ is real analytic on $\mathbf{T}^d$.*

(i) *If $v_0$ is a trigonometric polynomial or if $d = 1$, there is $\lambda_0 = \lambda_0(v_0)$ such that if $\lambda > \lambda_0$ and fixing some base point $x = x_0$, $H_\omega(x_0)$ satisfies Anderson localization for almost all $\omega$.*

(ii) *For $d$ arbitrary, specify a diophantine condition on $\omega, \omega \in DC_{A,c}$, i.e.,*

$$\|k.\omega\| > c|k|^{-A} \text{ for } k \in \mathbf{Z}^d \backslash \{0\}$$

*Then the preceding holds for $\lambda > \lambda_0(v_0, A, c)$ and almost all $\omega \in DC_{A,c}$.*

Observe that in case (i), the Lyapounov bound was indeed independent of $\omega$.

**Remarks.**

**1.** From [Bo] (see Chapter 7), part (i) of Corollary 10.34 remains valid for $d > 1$ with $\lambda > \lambda_0(v)$ and almost all $\omega$ (hence as nonperturbative result).

**2.** If $d = 1$, $v_0(x) = \cos x$, Herman's method (Proposition 3.1) implies the lower bound

$$L(E) > \log \frac{\lambda}{2} \text{ for any } E \in \mathbf{R}$$

and hence Corollary 10.34 applies with $\lambda > 2$. This is the localized regime in the Almost Mathieu model. In fact, S. Jitomirskaya proved an even more precise theorem under purely arithmetic conditions, implying in particular the localization property for any diophantine $\omega$ and taking $x = x_0$ in a set of measure 1. Her argument crucially uses the fact that

$$\det\{[\lambda \cos(\theta + n\omega)\delta_{nn'} + \Delta - E]_{|n|,|n'|<N}\}$$

is a polynomial in $\cos\theta$ of degree $2N - 1$. This is a special feature of the cosine-potential. The argument seems therefore restricted to this particular case.

**3.** For $d = 1$, results for diophantine $\omega$ were obtained by M. Goldstein and H. Eliasson in the perturbative regime (without explicit condition on $\lambda$). For $d = 2$, results in the spirit of Corollary 10.34(ii) were obtained by Chulaevsky and Sinai, again in the perturbative regime and under small measure-theoretic exclusion in the $\omega$-parameter set.

**4.** Taking the continuity properties of the Lyapounov exponent from Chapter 7 into account, a careful check of the proof of Theorem 10.1 shows that in fact the following is shown:

Fixing $x_0 \in \mathbf{T}^d$, there is a zero-measure set $\mathcal{R} \subset \mathbf{T}^d$ such that if $\omega \in DC \backslash \mathcal{R}$, then the following holds:

Let $\psi$ be an extended state of $H_\omega(x_0)$ (i.e., $\sup_n \frac{\log |\psi_n|}{\log(1+|n|)} < \infty$) with energy $E$ and $L(E) = L_\omega(E) > 0$. Then $E$ is in the point spectrum of $H_\omega(x_0)$, and $\psi$ is exponentially localized.

The proof of Theorem 10.1 also implies stronger results known as "dynamical localization." Consider the associated Schrödinger evolution

$$i\frac{\partial\psi}{\partial t} = H\psi \quad \psi(t) = e^{itH}\psi$$

where $\psi \in \ell^2(\mathbf{Z})$ is assumed, moreover, to satisfy a decay condition

$$|\psi_n| < |n|^{-A} \text{ for } |n| \to \infty \tag{10.35}$$

for some exponent $A$ large enough. Then, under the assumptions of Corollary 10.34, we may ensure that

$$\sup_t \left( \sum_{n \in \mathbf{Z}} (1 + n^2) |\langle e^{itH}\psi, \delta_n \rangle|^2 \right)^{1/2} < \infty \tag{10.36}$$

This property is referred to as "dynamical localization." It cannot hold if $H$ has a component of continuous spectrum.

Let $\{\varphi^{(\alpha)}, E^{(\alpha)}\}$ denote eigenstates and corresponding eigenvalues of $H$. Then

$$\psi = \sum_\alpha \langle \psi, \varphi^{(\alpha)} \rangle \varphi^{(\alpha)}$$

and hence

$$e^{itH}\psi = \sum_\alpha e^{itE^{(\alpha)}} \langle \psi, \varphi^{(\alpha)} \rangle \varphi^{(\alpha)}$$

It suffices, therefore, to estimate

$$\sum_\alpha \left( \sum_n (1 + n^2) |\varphi^{(\alpha)}(n)|^2 \right)^{1/2} |\langle \psi, \varphi^{(\alpha)} \rangle| \tag{10.37}$$

For simplicity, let $\psi = \delta_0$.

For each $\varepsilon > 0$, introduce the set

$$I_\varepsilon = \{\alpha| \ |\varphi^{(\alpha)}(0)| > \varepsilon\}$$

and estimate its size.

Returning to the proof of Theorem 10.1 and in particular estimate (10.13), the assumption $|\xi_0| > \varepsilon$ implies now that

$$\|G_{]-j_0,j_0[}(x_0, E_\alpha)\| > \varepsilon e^{\frac{c_0}{2}N_0}$$

(for some $0 < j_0 < N_0^C$). Thus we take $N_0 \gtrsim \log\frac{1}{\varepsilon}$, an exponential decay

$$|\varphi^{(\alpha)}(n)| < e^{-c|n|}$$

will start to appear for

$$|n| > N_\varepsilon \sim \left(\log\frac{1}{\varepsilon}\right)^{C'}$$

Observe that from a simple Hilbert-Schmidt argument

$$2N_\varepsilon + 1 \gg \sum_\alpha \|P_{[-N_\varepsilon,N_\varepsilon]}\varphi^{(\alpha)}\|^2 \geq \sum_{\alpha \in I_\varepsilon} \|P_{[-N_\varepsilon N_\varepsilon]}\varphi^{(\alpha)}\|^2 > \frac{1}{2}|I_\varepsilon|$$

so that

$$|I_\varepsilon| \lesssim N_\varepsilon$$

Moreover, for $\alpha \in I_\varepsilon$,

$$\left(\sum_n n^2 |\varphi^{(\alpha)}(n)|^2\right)^{1/2} \leq N_\varepsilon + \left(\sum_{|n|>N_\varepsilon} n^2 e^{-c|n|}\right)^{1/2} < N_\varepsilon + 1$$

Consequently,

$$(10.37) = \sum_\alpha \left(\sum_n (1+n^2)|\varphi^{(\alpha)}(n)|^2\right)^{1/2} |\varphi^{(\alpha)}(0)|$$

$$\lesssim \sum_{\varepsilon>0,\varepsilon \text{ dyadic}} \varepsilon.(N_\varepsilon + 1)|I_\varepsilon| < \sum_\varepsilon \varepsilon \left(\log\frac{1}{\varepsilon}\right)^{2C'} < \infty$$

Similar estimates may be ensured for $\psi = \delta_{n_0}$, $n_0$ arbitrary, with a powerlike bound in $|n_0|$, and hence permits us to establish (10.36) under an assumption (10.35).

# Chapter Eleven

## Generalization to Certain Long-Range Models

The preceding depends explicitly on the fundamental matrix formalism and hence requires nearest neighbor models (thus the off-diagonal is given by $\Delta$). To be precise, the proof of Proposition 7.19 for the Green's function is based on these techniques. Once this fact is established, our proof of localization would extend, for instance, more generally to Hamiltonians of the form

$$H = v(x + n\omega)\delta_{nn'} + S_\phi \qquad (11.1)$$

with $\Delta$ replaced by a Toeplitz operator $S_\phi$

$$S_\phi(n, n') = \hat{\phi}(n - n')$$

with $\phi$ real and decaying rapidly enough for $|n| \to \infty$.

Our purpose here is to establish (nonperturbative) localization results in the generality of (11.1). We will first treat the case of the cosine potential $v(x) = \cos x$, which is of special interest, as will be seen later.

**Theorem 11.2.** *Assume $\phi$ real analytic satisfying*

$$|\hat{\phi}(n)| < e^{-\rho|n|} \text{ for } n \in \mathbf{Z} \qquad (11.3)$$

*for some $\rho > 0$. Then there is $\varepsilon_0 = \varepsilon_0(\rho) > 0$ s.t. if $0 \le \varepsilon < \varepsilon_0$,*

$$H_\omega(x) = \cos(x + n\omega)\delta_{nn'} + \varepsilon S_\phi$$

*satisfi es A.L. for $(x, w) \in \mathbf{T}^2$ in a set of full measure.*

As was pointed out earlier, we only need to provide an analogue of Proposition 7.19. We will show the following.

**Proposition 11.4.** *Under the assumptions of Theorem 11.2, and taking $\omega$ diophantine, the following holds if $\varepsilon < \varepsilon_0(\rho)$:*

*Let $N$ be suffi ciently large. There is a subset $\Omega = \Omega_N(E) \subset \mathbf{T}$ satisfying*

$$\mathrm{mes}\,\Omega < e^{-N^\sigma}$$

*(for some $\sigma > 0$) such that if $\theta \notin \Omega$, then for some $m$, $|m| < N^{1/2}$ (and possibly dependent on $\theta$), we have the Green's function estimate*

$$|G_{[-N,N]}(E + i0, \theta + m\omega)(n, n')| < e^{-c(|n-n'|-\varepsilon_0^{\frac{1}{20}} N)}, \quad \text{for all } |n|, |n'| \le N. \qquad (11.5)$$

Notice that in the application to localization, a translation by $m = o(N)$ is harmless.

Before proving Proposition 11.4, we establish the following consequence of Proposition 11.4 and Aubry duality to continuous spectrum.

**Theorem 11.6.** *Let $\phi$ satisfy (11.3). Then, for $\lambda < \lambda_0(\rho)$ and $\omega$ diophantine,*

$$H_\omega = \lambda\phi(n\omega)\delta_{nn'} + \Delta \qquad (11.7)$$

*has only continuous spectrum.*

**Remark.** Theorem 11.6 as a *nonperturbative* result (i.e., $\lambda_0$ does not depend on $\omega$) holds only in the 1-frequency case ($\omega \in \mathbf{T}$). In the 2-frequency case, one typically may obtain some point spectrum for $\omega$ in a set of (small) positive measure, no matter how small $\lambda$.

In fact, Theorem 11.6 has the following strengthening (see [B-J2]), which will not be proven here.

**Theorem 11.8.** *Let $\phi$ satisfy (11.3). Then, for $\lambda < \lambda_0(\rho)$, $\omega$ diophantine and almost all $x$, $H_\omega(x) = \lambda\phi(x + n\omega)\delta_{nn'} + \Delta$ has purely absolutely continuous spectrum.*

To deduce Theorem 11.6, we use the Aubry duality reasoning in its simplest form. Assume $\psi$ an eigenvector of (11.7); hence $\psi \in \ell^2(\mathbf{Z})$ with eigenvalue $E$. Thus

$$(\lambda\phi(n\omega) - E)\psi_n + \psi_{n-1} + \psi_{n+1} = 0$$

Defining

$$F(\theta) = \sum_{n \in \mathbf{Z}} \psi_n e^{in\theta}$$

a function in $L^2(\mathbf{T})$ is obtained, satisfying the equation

$$\left(\cos\theta - \frac{E}{2}\right)F(\theta) + \frac{\lambda}{2}\sum_k \hat{\phi}(k)F(\theta + k\omega) = 0$$

For almost all $\theta$, the sequence

$$\xi_n = F(\theta + n\omega) \quad (n \in \mathbf{Z})$$

is well defined and will satisfy

$$\sum_{n \in \mathbf{Z}} \frac{|\xi_n|^2}{1 + n^2} < \infty \text{ hence } |\xi_n| < C|n|$$

and the equation

$$\left(\cos(\theta + n\omega) - \frac{E}{2}\right)\xi_n + \frac{\lambda}{2}\sum \hat{\phi}(k)\xi_{n+k} = 0$$

Thus, denoting

$$\widetilde{H}(\theta) = \cos(\theta + n\omega)\delta_{nn'} + \frac{\lambda}{2}S_\phi$$
$$\left(\widetilde{H}(\theta) - \frac{E}{2}\right)\xi = 0$$

and $\xi$ is an extended state of $\widetilde{H}(\theta)$.

Assume $\frac{|\lambda|}{2} < \varepsilon_0(\rho)$ for Proposition 11.4 to apply to $\widetilde{H}$. Fix $N$ and $\theta \notin \Omega_N(E)$ as above. Here $\Omega_N(E)$ is the set from Proposition 11.4.

Letting $\Lambda = [-N + m, N + m]$, $G_\Lambda(E + i0, \theta)$ satisfies (11.5). Write

$$R_\Lambda \xi = -G_\Lambda(E)(R_\Lambda S_{\frac{\lambda}{2}\phi} R_{\mathbf{Z} \backslash \Lambda})\xi$$

implying

$$|F(\theta)| = |\xi_0| < \sum_{\substack{n \in \Lambda \\ n' \notin \Lambda}} |G_\Lambda(E)(0, n)| \, |\hat{\phi}(n - n')| \, |\xi_{n'}|$$

$$< C \sum_{\substack{n \in \Lambda \\ n' \notin \Lambda}} e^{-c(|n| - \varepsilon_0^{\frac{1}{20}} N) - \rho|n - n'|}|n'|$$

$$< Ce^{-\min(c,\rho)\frac{N}{2}}$$

Letting $N \to \infty$, we conclude that $F = 0$ a.e., a contradiction. Therefore, (11.7) has no point spectrum.

**Remark.** For the Almost Mathieu operator

$$H_\omega^{(\lambda)}(x) = \lambda \cos(x + n\omega)\delta_{n,n'} + \Delta$$

the dual model is

$$\widetilde{H}_\omega(\theta) = \cos(\theta + n\omega)\delta_{n,n'} + \frac{\lambda}{4}\Delta = \frac{\lambda}{4} H_\omega^{\left(\frac{4}{\lambda}\right)}(\theta)$$

Since $\widetilde{H}$ satisfies AL for $\frac{4}{\lambda} > 2$, it follows that $H_\omega^{(\lambda)}$ has (absolutely) continuous spectrum for $\lambda < 2$.

**Proof of Proposition 11.4.** Let $H = H_\omega(x)$ be the Hamiltonian considered in Theorem 11.2. Estimate $G_{[0,N]}(E, x)$ by Cramer's rule

$$|G_{[0,N]}(E, x)(n, n')| = \frac{|\det A_{n,n'}(x)|}{|\det[H_N(x) - E]|} \tag{11.9}$$

where $A_{n,n'}(x)$ refers to the $(n, n')$-minor of $H_N(x) - E$.

Consider first the denominator $|\det[H_N(x) - E]|$ for which we need to establish a lower bound. The function

$$u(x) = \frac{1}{N} \log(|\det[H_N(x) - E]| + 10^{-N})$$

clearly admits a subharmonic extension to the complex plane, $\widetilde{u}(z)$, satisfying

$$-\log 10 < \widetilde{u}(z) < \log \left|\frac{1}{2}(e^{\operatorname{Im} z} + e^{-\operatorname{Im} z}) + \varepsilon \sum |\hat{\phi}(n)|\right|$$

Hence

$$|\hat{u}(k)| < \frac{C}{|k|}$$

Moreover, using Herman's trick,

$$\hat{u}(0) = \int_{\mathbf{T}} u > \frac{1}{N} \int \log \left|\det \left[\frac{1}{2}(e^{i\theta} + e^{-i\theta})\delta_{n,n'} + \varepsilon S_\phi - E\right]\right|$$

$$= \frac{1}{N} \int_{|z|=1} \log \left|\det \left[\frac{z^2 + 1}{2}\delta_{nn'} + z\varepsilon S_\phi - zE\right]\right|$$

$$\geq \frac{1}{N} \log \left|\det \left[\frac{1}{2}I\right]\right| = -\log 2 \tag{11.10}$$

The function $u = u(x)$ is not invariant under the shift $x \mapsto x + \omega$. But we may define for $R = o(N)$ the function

$$v(x) = \sum_{0 \leq |j| < R} \frac{R - |j|}{R^2} u(x + j\omega),$$

for which

$$\langle v \rangle = \hat{u}(0) \text{ and } |v(x + \omega) - v(x)| \lesssim \frac{1}{R}$$

Assuming $\omega$ diophantine and returning to the proof of Theorem 5.1 permit us to show (taking $R = \sqrt{N}$) that

$$\text{mes}\,[x \in \mathbf{T}\,|\,|v(x) - \hat{u}(0)| > N^{-\sigma}] < e^{-N^\sigma}$$

for some $\sigma > 0$. Thus, outside a set $\Omega = \Omega_N(E)$, $\text{mes}\,\Omega < e^{-N^\sigma}$, we get

$$v(x) > \hat{u}(0) - N^{-\sigma} > -\log 2 - N^{-\sigma}$$

Equivalently, for $x \notin \Omega$,

$$\max_{|m|<\sqrt{N}} |\det[H_N(x + m\omega) - E]| > 2^{-N-N^{1-\sigma}} \tag{11.11}$$

It remains to obtain an upper bound on $|\det A_{nn'}(x)|$. This is achieved by expressing $\det A_{nn'}$ as a sum over "paths" $\gamma$ as follows:

$$A_{nn'} = \sum_s \sum_{\gamma,|\gamma|=s} \pm(\det[R_{[0,N[\backslash\gamma}(H - E)R_{[0,N[\backslash\gamma}])\varepsilon^{s-1}\prod_{i=1}^{s-1}\hat{\phi}(\gamma_{i+1} - \gamma_i) \tag{11.12}$$

where $\gamma = (\gamma_1,\ldots,\gamma_s)$ is a sequence in $[0, N[$ with $\gamma_1 = n$, $\gamma_s = n'$.
Hence

$$|A_{nn'}(x)| < \sum_s \sum_{\gamma,|\gamma|=s} \varepsilon^{s-1}e^{-\rho\sum_{i=1}^{s-1}|\gamma_{i+1}-\gamma_i|}\,|\det[R_{[0,N[\backslash\gamma}(H(x) - E)R_{[0,N[\backslash\gamma}]|$$

$$< \sum_{b\geq|n-n'|}\sum_{s\leq b} 2^{s-1}\binom{b}{s-1}\varepsilon^{s-1}e^{-\rho b}\max_{\gamma(s,b)-\text{path}}|\det[R_{\gamma c}(H(x) - E)R_{\gamma c}]| \tag{11.13}$$

where we denoted

$$b \equiv \sum_{i=1}^{s-1}|\gamma_{i+1} - \gamma_i| \geq |n - n'|$$

and used the fact that there are at most $2^{s-1}\binom{b}{s-1}$ $(s,b)$-paths.

Estimating the determinant by Hadamard, we find

$$|\det[R_{\gamma c}(H(x) - E)R_{\gamma c}]|$$
$$\leq \prod_{k\in[0,N[\backslash\gamma}\left[|\cos(x + k\omega) - E + \varepsilon\hat{\phi}(0)|^2 + \varepsilon^2\sum_{j\neq 0}|\hat{\phi}(j)|^2\right]^{1/2} \tag{11.14}$$

The estimate on (11.14) and $|A_{nn'}(x)|$ will be uniformly in $x$.
Write

$$\log(11.14) \leq \sum_{k\in[0,N[\backslash\gamma}\log[|\cos(x + k\omega) - E| + \varepsilon\|\hat{\phi}\|_1]$$

We clearly may assume $|E| < 1 + \varepsilon_0\|\hat{\phi}\|_1$ and then obtain a bound

$$\sum_{k\in[0,N[}\log[|\cos(x + k\omega) - E_0| + 2\varepsilon_0(1 + \|\hat{\phi}\|_1)] \tag{11.15}$$

$$-\sum_{k\in\gamma}\log[|\cos(x+k\omega)-E_0|+2\varepsilon_0(1+\|\hat{\phi}\|_1)] \qquad (11.16)$$

where $E_0=\cos x_0$ for some $x_0$.

Majorize (for $N$ large enough)

$$(11.15)\leq N\int_{\mathbf{T}}\log[|\cos x-E_0|+2\varepsilon_0(1+\|\hat{\phi}\|_1)]+N^{1-\delta}$$

$$\leq N\int_{\mathbf{T}}\log|\cos x-E_0|+\varepsilon_0^{\frac{1}{2}-}(1+\|\hat{\phi}\|_1)N+N^{1-\delta}$$

$$<-N\log 2+\varepsilon_0^{\frac{1}{2}-}N$$

Here $\delta=\delta(\omega)>0$. In the second inequality, use the fact that

$$\int_{\mathbf{T}}|\cos x-E_0|^{\frac{1}{2}-}<C$$

Next, a minoration on (10.16) is needed.

There is always the obvious lower bound

$$\sum_{k\in\gamma}\log[|\cos(x+k\omega)-E_0|+2\varepsilon_0(1+\|\hat{\phi}\|_1)]>s\log\varepsilon_0$$

where

$$s=|\gamma|$$

Assume $s>\varepsilon_0^{\frac{1}{10}}N$. Then a better lower bound will be given.

Since $\omega$ in DC, we have for $N$ large enough, $\kappa>\frac{1}{\log N}$, that

$$\#\{k=0,1,\ldots,N-1|\ \|k\omega+x\pm\theta_0\|<\kappa\}<10\kappa N$$

Letting $\kappa\sim\frac{s}{N}$, it follows that for at least $\frac{s}{2}$ elements $k\in\gamma$

$$\log[\cdots]>\log\kappa^2>\log 10^{-3}\left(\frac{s}{N}\right)^2>\log\varepsilon_0^{1/4}$$

and thus

$$\sum_{k\in\gamma}\log[\cdots]>\frac{s}{2}\log\varepsilon_0+\frac{s}{2}\log\varepsilon_0^{1/4}>\frac{3}{4}s\log\varepsilon_0$$

In summary,

$$\log(11.14)<-(\log 2)N+\varepsilon_0^{\frac{1}{2}-}N+s.\log\frac{1}{\varepsilon_0} \qquad (11.17)$$

and if $s>\varepsilon_0^{\frac{1}{10}}N$,

$$\log(11.14)<-(\log 2)N+\varepsilon_0^{\frac{1}{2}-}N+\frac{3}{4}s\log\frac{1}{\varepsilon_0} \qquad (11.18)$$

Returning to (11.13), we obtain from the preceding

$$|A_{nn'}|<$$

$$2^{-N}\sum_{b\geq|n-n'|}\sum_{\substack{s\leq b\\ s\leq\varepsilon_0^{\frac{1}{10}}}}2^s\binom{b}{s-1}\varepsilon^{s-1}e^{-\rho b}e^{\varepsilon_0^{\frac{1}{2}-}N}\left(\frac{1}{\varepsilon_0}\right)^s$$

$$+2^{-N}\sum_{b\geq|n-n'|}\sum_{\substack{s\leq b\\ s>\varepsilon_0^{\frac{1}{10}}N}}2^s\binom{b}{s-1}\varepsilon^{s-1}e^{-\rho b}e^{\varepsilon_0^{\frac{1}{2}-}N}\left(\frac{1}{\varepsilon_0}\right)^{\frac{3}{4}s}$$

and distinguishing the cases $b \leq \varepsilon_0^{\frac{1}{20}} N, b \geq \varepsilon_0^{\frac{1}{20}} N$, for $\varepsilon_0 = \varepsilon_0(\rho)$ small enough

$$< 2^{-N} e^{\varepsilon_0^{\frac{1}{2}} - N} (2^{\varepsilon_0^{\frac{1}{20}} N} . e^{-\rho|n-n'|} + e^{-\frac{\rho}{2}|n-n'|}) \tag{11.19}$$

From (11.11) and (11.19), it results that for $x \notin \Omega_N(E)$, there is some $m, |m| < \sqrt{N}$ for which

$$|G_N(E + i0, x + m\omega)| < e^{N^{1-\sigma} + \varepsilon_0^{\frac{1}{20}} N - \frac{\rho}{2}|n-n'|}$$

Thus (11.5) holds. This proves Proposition 11.4.

Theorem 11.2 generalizes to

**Theorem 11.20.** *Let $\phi$ and $v$ be real analytic on $\mathbf{T}$, $v$ nonconstant. Then, for $0 < \varepsilon < \varepsilon_0, \varepsilon_0 = \varepsilon_0(v, \phi)$,*

$$H_\omega(x) = v(x + n\omega)\delta_{nn'} + \varepsilon S_\phi$$

*satisfies A.L. for $(x, \omega) \in \mathbf{T}^2$ in a set of full measure.*

Again, the issue is Proposition 11.4, under the assumptions of Theorem 11.20. Analyzing the proof of Proposition 11.4, the main problem is to obtain the "correct" lower bound on $\frac{1}{N} \int_{\mathbf{T}} \log |\det[H_N(x) - E]|dx$. In the special case $v(x) = \cos x$, this is achieved by Herman's argument. In other situations (with $v(x)$ given by a trigonometric polynomial), that argument would not provide the right estimate. Using a variant on the proof of Proposition 3.3, we show the following:

**Proposition 11.21.** *Under the assumptions of Theorem 11.20, we have for diophantine $\omega$*

$$\frac{1}{N} \int_{\mathbf{T}} \log |\det[H_N(x) - E]|dx > \int_{\mathbf{T}} \log |v(x) - E|dx - \kappa(\varepsilon) \tag{11.22}$$

*where $\kappa(\varepsilon) \to 0$ for $\varepsilon \to 0$ and taking $N$ large enough.*

**Proof.** Assume $|\hat{v}(k)| < e^{-|k|\rho} (k \in \mathbf{Z}), \rho > 0$. Since $v$ is nonconstant,

$$\min_E \int_0^1 \log |v(x) - E|dx > -C_v \tag{11.23}$$

for some constant $C_v$.

Fix $0 < \sigma < \frac{\rho}{2}$. Proceeding as in the proof of Proposition 3.3, there is $\delta_0 > 0$ s.t.

$$\min_E \max_{\frac{\sigma}{2} < y_0 < \sigma} \min_{x \in \mathbf{R}} |v(x \pm iy_0) - E| > \delta_0 \tag{11.24}$$

Take

$$\varepsilon_0 = \delta_0^2, \quad 0 < \varepsilon < \varepsilon_0$$

For $z = x + iy, |y| < \frac{\rho}{2}$, estimate by Hadamard

$$|\det[H_N(z) - E]| \leq \prod_{0 \leq n < N} (|v(z + n\omega) - E| + C\varepsilon_0)$$

$$\frac{1}{N} \log |\det[H_N(z) - E]| \leq \frac{1}{N} \sum_{n=0}^{N-1} \log[|v(x + iy + n\omega) - E| + C\varepsilon_0]$$

$$= \int_0^1 \log[|v(x + iy) - E| + C\varepsilon_0]dx + 0(N^{-\delta_1})$$

$$< C \tag{11.25}$$

Here $\delta_1 = \delta_1(\omega)$, $\omega$ is diophantine, and $N$ assumed large enough. We may and do assume $E$ bounded.

Next, consider the diagonal matrix $D$

$$D_{nn} = v(x + iy_0 + n\omega) - E \quad (0 \le n < N)$$

and with $y_0$ as in (11.24). Thus

$$H_N(x + iy_0) - E = D_{n,n}\delta_{nn'} + \varepsilon S_\phi^{(N)}$$

and writing

$$H_N(x + iy_0) - E = \left(H_N(x + iy_0) - E\right)D^{-1}D$$

it follows that

$$\frac{1}{N} \log |\det[H_N(x + iy_0) - E]| = \frac{1}{N} \sum_{n=0}^{N-1} \log |D_{nn}| \qquad (11.26)$$
$$+ \frac{1}{N} \log |\det[I + \varepsilon S_\phi^{(N)} D^{-1}]| \qquad (11.27)$$

where, by (11.24)

$$\|D^{-1}\| < \frac{1}{\delta_0} < \varepsilon^{-1/2}$$

Write

$$(11.26) \ge \frac{1}{N} \sum_{n=0}^{N-1} \log \left(|v(x + iy_0 + n\omega) - E| + \varepsilon_0\right) - \frac{\varepsilon_0}{\delta_0}$$

$$= \int_0^1 \log(|v(x + iy_0) - E| + \varepsilon_0)dx - \frac{\varepsilon_0}{\delta_0} + 0(N^{-\delta_1})$$

$$> \int_0^1 \log |v(x + iy_0) - E|dx - 2\frac{\varepsilon_0}{\delta_0}$$

Write next

$$\log |\det[I + \varepsilon S_\phi D^{-1}]| = -\log |\det[I + \varepsilon S_\phi D^{-1}]^{-1}|$$

$$= -\log |\det[I + \sum_{s \ge 1}(-1)^s(\varepsilon S_\phi D^{-1})^s]|$$

$$\ge -\log \prod_{n=0}^{N-1} \left[1 + \sum_{s \ge 1} \varepsilon^s \|(S_\phi D^{-1})^s e_n\|\right]$$

$$\ge -\sum_{n=0}^{N-1} \log(1 + 2\varepsilon \|S_\phi D^{-1} e_n\|)$$

$$\ge -\sum_{n=0}^{N-1} \log(1 + C\varepsilon |D_{nn}^{-1}|)$$

$$> -CN\varepsilon\delta_0^{-1}$$

Therefore, for all $x \in \mathbf{R}$,

$$\frac{1}{N} \log |\det[H_N(x + iy_0) - E]| > \int_0^1 \log |v(x + iy_0) - E|dx - C\frac{\varepsilon_0}{\delta_0} \quad (11.28)$$

Next, we exploit subharmonicity of the function

$$u(z) = \frac{1}{N} \log |\det[H_N(z) - E]| \quad (|\text{Im } z| < \rho)$$

Thus, fixing $x$, we get with $y_1 = \frac{\rho}{2}$,

$$u(x + iy_0) \leq \int_{y=0} u(x')\mu_{x+iy_0}(dx') + \int_{y=y_1} u(x' + iy_1)\mu_{x+iy_0}(dx')$$

$$= \int_{y=0} u(x + x')\mu_{iy_0}(dx') + \int_{y=y_1} \mu(x + x' + iy_1)\mu_{iy_0}(dx')$$

where $\mu_z \in \mathcal{M}([y = 0] \cup [y = y_1])$ is the harmonic measure of $z$ in the strip $0 \leq \text{Im } z \leq y_1$.

Averaging in $x$ implies

$$\int_0^1 u(x + iy_0)dx \leq \left[\int_0^1 u(x)dx\right]\mu_{iy_0}[y = 0] + \left[\int_0^1 u(x + iy_1)dx\right]\mu_{iy_0}[y = y_1]$$

$$= \left(\int_0^1 u(x)dx\right)\left(1 - \frac{y_0}{y_1}\right) + \left(\int_0^1 u(x + iy_1)dx\right)\frac{y_0}{y_1}$$

and recalling (11.28) and (11.25),

$$\int_0^1 \log|v(x + iy_0) - E|dx - C\frac{\varepsilon_0}{\delta_0} < \left(1 - \frac{y_0}{y_1}\right)\left(\int_0^1 u(x)dx\right) + C\frac{y_0}{y_1}$$

Therefore, recalling also (11.23),

$$\int_0^1 u(x)dx > \left(1 + \frac{y_0}{y_1 - y_0}\right)\int_0^1 \log|v(x + iy_0) - E|dx - C\left(\frac{\varepsilon_0}{\delta_0} + \frac{\sigma}{\rho}\right)$$

$$> \int_0^1 \log|v(x + iy_0) - E|dx - C_v\frac{\sigma}{\rho} - C\varepsilon_0^{1/2}$$

Subject to replacement of $y_0$ by $-y_0$, (see (11.24)), and since

$$\int_0^1 \log|v(x) - E|dx \leq \frac{1}{2}\int \log|v(x + iy_0) - E|dx + \frac{1}{2}\int \log|v(x - iy_0) - E|dx$$

it follows that

$$\frac{1}{N}\int_0^1 \log|\det[H_N(x) - E]|dx > \int \log|v(x) - E|dx - C\left(\frac{\sigma}{\rho} + \varepsilon_0^{1/2}\right)$$

This proves Proposition 11.21.

In order to get an upper bound on

$$|\det[R_{\gamma c}(H_N(x) - E)R_{\gamma c}]|$$

where $\gamma \subset \{0, 1, \ldots, N - 1\}$, we use the following.

**Lemma 11.29.** If $\gamma \subset \{0, 1, \ldots, N - 1\}$ and $|\gamma| = s > \varepsilon_1(\varepsilon_0)N$ where $\log\frac{1}{\varepsilon_0} \sim \log\frac{1}{\varepsilon_1}$, then for all $x$

$$\sum_{k \in \gamma} \log[|v(x + k\omega) - E| + \varepsilon_0] > \frac{3}{4}s\log\varepsilon_0 \tag{11.30}$$

**Proof.** Approximation by trigonometric polynomials permits for fixed $\kappa$ to get $w(x) = \sum_{|k|<d} \hat{w}(k)e^{ikx}, d < C \log \frac{1}{\kappa}$, such that $\|v - w\|_\infty < \kappa$. Hence

$$[x \in [0,1] \mid |v(x) - E| < \kappa] \subset [x \in [0,1] \mid |w(x) - E| < 2\kappa]$$
$$\subset [x \in [0,1] \mid |v(x) - E| < 3\kappa]$$

is contained in a union of at most $Cd^2$ intervals of total measure at most $\kappa^c$, $c = c_v > 0$. Thus for $N$ large enough

$$\#\{k = 0, 1, \ldots, N-1 \mid |v(x + k\omega) - E| < \kappa\} < 2\kappa^c N$$

and appropriate choice of $\kappa$ implies

$$\sum_{k \in \gamma} \log[|v(x + k\omega) - E| + \varepsilon_0] > \frac{s}{2} \log \varepsilon_0 + C\frac{s}{2} \log \frac{s}{N}$$

The lemma follows.

**Remarks.**
**1.** Proposition 11.21 implies the following refinement of Proposition 3.3 (assuming $\omega$ diophantine).

**Proposition 11.31.** *Let* $H(x) = \lambda v(x + n\omega)\delta_{nn'} + \Delta$ *(1-frequency on* **Z***) with* $v$ *nonconstant real analytic and* $\omega$ *diophantine. Then, for* $\lambda > \lambda_0(v)$*, we have*

$$L(E) > \int_{\mathbf{T}} \log|\lambda v(x) - E| dx - \kappa(\lambda)$$

*where* $\kappa(\lambda) \to 0$ *for* $\lambda \to \infty$*.*
**2.** The method described above permits us also to establish localization results for *band-Schrödinger operators* with index set

$$\mathbf{Z} \times \{1, \ldots, b\} \tag{11.32}$$

Thus, let $H(\omega, \theta)$ be the following lattice Schrödinger operator on (11.32):

$$H_{(n,s),(n',s')}(\omega, \theta) = \lambda v_s(\theta + n\omega)\delta_{nn'}\delta_{ss'} + \Delta \tag{11.33}$$

where $\{v_s | s = 1, \ldots, b\}$ are real analytic, nonconstant on $\mathbf{T}$, and $\Delta$ stands for the Laplacian on $\mathbf{Z}^2$

$$\Delta((n,s),(n',s')) = \begin{cases} 1 \text{ if } |n - n'| + |s - s'| = 1 \\ = 0 \text{ otherwise} \end{cases}$$

**Theorem 11.34.** *Consider* $H$ *as above. Then for* $\lambda > \lambda_0(v_1, \ldots, v_b)$*, Anderson localization holds for* $(\omega, \theta) \in \mathbf{T}^2$ *in a set of full measure.*
See [B-J1].

**Remark.** In the case of quasi-periodic Schrödinger operators on $\mathbf{Z}^2$

$$H(\omega_1, \omega_2, ; \theta_1, \theta_2) = \lambda v(\theta_1 + n_1\omega_1, \theta_2 + n_2\omega_2) + \Delta$$

($v$ real analytic on $\mathbf{T}^2$), only perturbative localization results may be expected (see [BGS] and Chapter 17).

## References

[B-J1] J. Bourgain, S. Jitomirskaya. Anderson localization for the band model, *Springer LNM* 1745 (2000), 67–79.

[B-J2] Idem. Non-perturbative absolutely continuous spectrum for 1D quasi-periodic operators, *Inventiones Math.* 148(3) (2002), 453–463.

[BGS] J. Bourgain, M. Goldstein, W. Schlag. Anderson localization on $\mathbf{Z}^2$ for Schrödinger operators with quasi-periodic potential, *Acta Math.* 188 (2002), 41–86.

# Chapter Twelve

## Lyapounov Exponent and Spectrum

First, we recall some basic facts from spectral theory.

Let $H$ be a bounded self-adjoint operator on $\ell^2(\mathbf{Z})$. Then, for $z \in \mathbf{C}\backslash\operatorname{Spec} H$, $(H-z)^{-1}$ is analytic (hence in particular for $\operatorname{Im} z > 0$), and we have for $f \in \ell^2$

$$\operatorname{Im}\langle(H-z)^{-1}f, f\rangle = \operatorname{Im} z.\|(H-z)^{-1}f\|^2 \qquad (12.1)$$

Thus

$$\phi_f(z) = \langle(H-z)^{-1}f, f\rangle$$

is an analytic function on the upper half plane with $\operatorname{Im}\phi_f \geq 0$ ($\phi_f$ is a so-called Herglotz function).

Therefore, one has a representation

$$\phi_f(z) = \langle(H-z)^{-1}f, f\rangle = \int_{\mathbf{R}} \frac{1}{\lambda - z}\mu_f(d\lambda)$$

where $\mu_f$ is the spectral measure associated to $f$. Thus $\mu_f \geq 0, \|\mu_f\| = \|f\|^2$.

If for $f, g \in \ell^2$ we let

$$\mu_{f,g} = \frac{1}{4}[\mu_{f+g} - \mu_{f-g} + i(\mu_{f+ig} - \mu_{f-ig})]$$

then

$$\langle(H-z)^{-1}f, g\rangle = \int_{\mathbf{R}} \frac{1}{\lambda - z}\mu_{f,g}(d\lambda)$$

Decompose

$$\mu_f = \mu_{f,pp} + \mu_{f,sc} + \mu_{f,ac}$$

is discrete, singular continuous, and absolutely continuous parts,

Define $\sum_{pp} = \sum_{pp}(H) = $ complement maximal open set $G$ in $\mathbf{R}$ for which $\mu_{f,pp}(G) = 0, \forall f$ and similarly $\sum_{sc}, \sum_{ac}$. Let $\sum = \sum_{pp} \cup \sum_{sc} \cup \sum_{ac} = $ $\operatorname{Spec} H = \{E \in \mathbf{R}\,|\inf_{\|f\|=1}\|(H-E)f\| = 0\}$. Next, we will recall some facts from Kotani's theory, such as the Ishii-Pastur-Kotani theorem. We consider the context of Schrödinger Hamiltonians on $\mathbf{Z}$ associated with a shift

$$H(x) = v(x + n\omega)\delta_{nn'} + \Delta \qquad (12.2)$$

($\omega$ DC), as studied earlier. The results mentioned below apply in much greater generality, however. There are also several other important facts from this theory that will not be brought up here.

Denoting again $L(E)$ the Lyapounov exponent, recall that in the context (12.2), $L(E)$ is a continuous function of $E$, and in particular, $[E \in \mathbf{R}\,|L(E) = 0]$ is closed.

**Proposition 12.3.** *(Pastur-Ishii)*

$$[L(E) > 0] \cap \sum_{ac}^{x} = \phi \text{ hence } \sum_{ac}^{x} \subset [L(E) = 0]$$

*x almost surely. (Notice that* $\sum_{pp}, \sum_{ac}, \sum_{sc}$ *do depend on* $x$.)

**Proof of Proposition 12.3.** Consider the set

$$\mathcal{E}_x = \{E \in \mathbf{R} \mid H(x) \text{ has a generalized eigenstate } \xi \text{ with eigenvalue } E\}$$

This means that

$$(H(x) - E)\xi = 0 \tag{12.4}$$

where, say,

$$|\xi_n| < |n| \text{ for } |n| \to \infty$$

It is well known that for $f \in \ell^2$,

$$\mu_f^x(\mathcal{E}_x) = 1$$

On the other hand, if $L(E) > 0$, then the Green's function $G_{[-N,N]}(E + io, x)$ satisfies an estimate

$$|G_{[-N,N]}(E + io, x)(n, n')| < e^{-c|n-n'|} \text{ for } |n - n'| > \frac{N}{10}$$

except for $x \in \Omega_N(E)$, $\text{mes}\,\Omega_N(E) < \frac{1}{N^2}$, for $N$ large.

Assume (12.4).

Fix $j_0$ s.t. $\xi_{j_0} \neq 0$, and let $N > 2j_0$. If $x \notin \Omega_N(E)$, we get

$$0 < |\xi_{j_0}| \leq |G_N(E, x)(j_0, N)|\,|\xi_{N+1}| + |G_N(E, x)(j_0, -N)|\,|\xi_{-N-1}|$$
$$< e^{-cN}(|\xi_{N+1}| + |\xi_{-N-1}|)$$

In particular, for $L(E) > 0$,

$$x \notin \overline{\lim}\,\Omega_N(E) \Rightarrow E \notin \mathcal{E}_x$$

and thus

$$\text{mes}\,[(x, E) \in \mathbf{T} \times \mathbf{R} \mid L(E) > 0 \text{ and } E \in \mathcal{E}_x] = 0$$

Write

$$\int_{\mathbf{T}} \mu_{f,ac}^x[E \mid L(E) > 0]dx = \sup_{K>0} \int_{\mathbf{T}} \mu_{f,ac}^x\left[E \mid L(E) > 0, \left|\frac{d\mu_{f,ac}^x}{dE}\right| < K\right]$$

where, by the preceding, the integral is

$$\leq K \int_{\mathbf{T}} \text{mes}\,[E \mid L(E) > 0, E \in \mathcal{E}_x]dx = 0$$

Hence, $x$ almost surely

$$\mu_{f,ac}^x[E \mid L(E) > 0] = 0$$

proving the result.

Proposition 12.3 has a converse due to Kotani.

**Proposition 12.5.** *(Kotani) Let $I \subset \mathbf{R}$ be an interval s.t. $\mathrm{mes}(I \cap [L(E) = 0]) > 0$. Then $\sum_{ac}^x \cap I \neq \phi$, $x$ a.s.*

**Corollary 12.6.**

$$\mathrm{mes}\left([L(E) = 0] \backslash \sum_{ac}^x\right) = 0, \quad x \text{ a.s.}$$

**Proof of Proposition 12.5.** We follow basically the exposition in [Si]. Let $\mathrm{Im}\, z > 0$. Define

$$u_\pm = (H_\pm - z)^{-1} e_0 \in \ell^2(\mathbf{Z}_\pm \cup \{0\})$$

Here $H_+$ (resp. $H_-$) is the restriction of $H$ to $[0, \infty[$ (resp. $]-\infty, 0]$).
It follows that

$$(H - z)\big(u_-(0)u_+ + u_+(0)u_- - u_+(0)u_-(0)e_0\big)$$
$$= [u_-(0)u_+(1) + u_+(0)u_-(-1) + \big(v(x) - z\big)u_+(0)u_-(0)]e_0$$

and hence

$$u_+(0)u_-(0)$$
$$= G(z, x)(0, 0)[u_-(0)u_+(1) + u_+(0)u_-(-1) + \big(v(x) - z\big)u_+(0)u_-(0)]$$

Since

$$u_\pm(0) = \langle (H_\pm - z)^{-1} e_0, e_0 \rangle \neq 0$$

we may divide by $u_+(0).u_-(0)$ and obtain, denoting

$$m_\pm(x, z) = -\frac{u_\pm(\pm 1)}{u_\pm(0)}$$

that

$$G(z, x)(0, 0)^{-1} = -m_+(x, z) - m_-(x, z) + v(x) - z \qquad (12.7)$$

Denote $H'_+$ (resp. $H'_-$) the restriction of $H$ to $]0, \infty[$ (resp. $]-\infty, 0[$). Since

$$(H'_+ - z)\left(\sum_{n \geq 1} u_+(n)e_n\right) = [(v(x + w) - z)u_+(1) + u_+(2)]e_1 = -u_+(0)e_1$$

and hence

$$u_+(1) = -\langle (H'_+ - z)^{-1} e_1, e_1 \rangle u_+(0)$$
$$m_+(x, z) = \langle (H'_+(x) - z)^{-1} e_1, e_1 \rangle$$

it follows that $m_\pm(x, z)$ are Herglotz functions of $z$.
Let $\tilde{u}_+$ be the extension of $u_+$ to $\mathbf{Z}$ satisfying

$$(H - z)\tilde{u}_+ = 0$$

(thus $\tilde{u}_+(n) = u_+(n)$ for $n \geq 0$ but $\tilde{u}_+ \notin \ell^2$).
Taking covariance considerations into account, it follows that

$$\tilde{u}_+^{(x)} \sim S\tilde{u}_+^{(Tx)}$$

where

$$Tx = x + w \text{ and } (S\xi)_n = \xi_{n-1}$$

Since

$$\left(v(x) - z\right) + \frac{\tilde{u}_+(1)}{\tilde{u}_+(0)} + \frac{1}{\frac{\tilde{u}_+(0)}{\tilde{u}_+(-1)}} = 0$$

we obtain therefore

$$v(x) - z - m_+(x, z) - \frac{1}{m_+(T^{-1}x, z)} = 0 \qquad (12.8)$$

Taking imaginary parts, it follows that

$$\operatorname{Im} z + \operatorname{Im} m_+(x, z) = \frac{\operatorname{Im} m_+(T^{-1}x, z)}{|m_+(T^{-1}x, z)|^2}$$

$$2 \log |m_+(T^{-1}x, z)| = \log \operatorname{Im} m_+(T^{-1}x, z) - \log[\operatorname{Im} z + \operatorname{Im} m_+(x, z)]$$

and integrating in $x$

$$2 \int \log |m_+(x, z)| dx = - \int \log \left(1 + \frac{\operatorname{Im} z}{\operatorname{Im} m_+(x, z)}\right) dx$$

Applying the inequality

$$\log(1 + t) \geq \frac{t}{1 + t} \text{ for } t \geq 0$$

we get

$$2 \int \log |m_+(x, z)| dx \leq - \int \frac{\operatorname{Im} z}{\operatorname{Im} m_+(x, z) + \operatorname{Im} z} dx \qquad (12.9)$$

Write next

$$\frac{u_n}{u_0} = \frac{u_n}{u_{n-1}} \cdot \frac{u_{n-1}}{u_{n-2}} \cdots \frac{u_1}{u_0}$$

$$\sum_{j=1}^{n} \log |m_+(T^{j-1}x, z)| = \sum_{j=1}^{n} \log \left|\frac{u_j}{u_{j-1}}\right| = \log \left|\frac{u_n}{u_0}\right|$$

and

$$\begin{pmatrix} \frac{u_n}{u_0} \\ \frac{u_{n+1}}{u_0} \end{pmatrix} = M_n(x, z) \begin{pmatrix} 1 \\ \frac{u_1}{u_0} \end{pmatrix}$$

Hence

$$1 \leq \|M_n(x, z)^{-1}\| \left(\frac{|u_n|}{|u_0|} + \frac{|u_{n+1}|}{u_0}\right)$$

$$- \log \|M_n(x, z)\| \leq \log \left(\left|\frac{u_n}{u_0}\right| + \left|\frac{u_{n+1}}{u_0}\right|\right)$$

Dividing by $n$ and letting $n \to \infty$, it follows that

$$\int \log |m_+(x, z)| dx \geq - \lim \int \frac{1}{n} \log \|M_n(x, z)\| dx = -L(z) \qquad (12.10)$$

where $L(z)$, $z \in \mathbf{C}$ is the subharmonic extension of the Lyapounov exponent satisfying $L(z) < \log(|z| + C)$.

Thus, from (12.9) and (12.10),

$$\int \frac{dx}{\operatorname{Im} m_+(x,z) + \operatorname{Im} z} \leq 2\frac{L(z)}{\operatorname{Im} z} \tag{12.11}$$

By assumption, $K = [L(E) = 0] \cap I$ has positive measure.
We may write on a complex neighborhood of $I$

$$L(z) = \int \log|z - w|\rho(dw)$$

where $\rho$ is some positive measure. Thus

$$\lim_{\varepsilon \to 0} \frac{L(E + i\varepsilon) - L(E)}{\varepsilon}$$

exists for almost all $E$, and hence

$$\lim_{\varepsilon \to 0} \frac{L(E + i\varepsilon)}{\varepsilon}$$

exists for almost all $E \in K$.
In fact, taking $z = E + i\varepsilon, \varepsilon > 0$ in (12.11), one has that

$$\overline{\lim_{\varepsilon > 0}} \int_K \int_{\mathbf{T}} \frac{dx dE}{\operatorname{Im} m_+(x, E + i\varepsilon) + \varepsilon} < \infty \tag{12.12}$$

Returning to (12.7) and taking imaginary parts,

$$\operatorname{Im} m_+(x, E + i\varepsilon) + \operatorname{Im} m_-(x, E + i\varepsilon) + \varepsilon = \frac{\operatorname{Im} G(E + i\varepsilon, x)(0,0)}{|G(E + i\varepsilon, x)(0,0)|^2}$$

where $\operatorname{Im} m_+ \geq 0, \operatorname{Im} m_- \geq 0$. Therefore, (12.12) implies $x$ a.s.

$$\lim_{\varepsilon} \int_K \frac{|G(E + i\varepsilon, x)(0,0)|^2}{[\operatorname{Im} G(E + i\varepsilon, x)(0,0)]} < \infty \tag{12.13}$$

Assume that the spectral measure $\mu_{\{0\}}^x$ has no a.c. component in $I$. Then

$$\lim_{\varepsilon \to 0} \operatorname{Im} G(E + i\varepsilon, x)(0,0) = 0 \quad E \text{ a.s. on } K$$

and (12.13) implies

$$\lim_{\varepsilon \to 0} G(E + i\varepsilon, x)(0,0) = 0 \quad \text{a.e. on } K$$

This is a contradiction, however, because $G(z, x)(0,0)$ is Herglotz and mes $K > 0$.
Proposition 12.5 is proven.
**Reference**

[Si] B. Simon. Kotani theory for one dimensional stochastic Jacobi matrices, *CMP* 89 (1983), 227–234.

Since from the covariance property

$$H(x + \omega) = S^{-1}H(x)S \quad (S = \text{shift})$$

it follows that

$$\operatorname{Spec} H(x) = \operatorname{Spec} H(x + \omega)$$

and hence

$$\operatorname{Spec} H = \operatorname{Spec} H(x) = \sum\nolimits^x = \sum\nolimits^x_{pp} \cup \sum\nolimits^x_{sc} \cup \sum\nolimits^x_{ac}$$

does not depend on $x$.

Notice that each of the components may depend on $x$, however. For instance, in the Almost Mathieu case $H(x) = \lambda \cos(x + n\omega)\delta_{nn'} + \Delta$ with $\lambda$ large and $\omega DC$, $\sum^x_{pp} = \phi$ for an uncountable (measure 0) set of $x$-values.

In what follows we continue to consider operators $H(x) = \lambda v(x + n\omega)\delta_{nn'} + \Delta$, $\omega DC$.

**Proposition 12.14.** *(Bourgain) Assume that $L(E) > 0$ for all energies $E$. Then Spec $H$ has positive Lebesgue measure.*

We treat the 1-frequency case. There are some additional technicalities to deal with the multifrequency case we don't want to go into here. The argument is also generalizable to other models.

**Lemma 12.15.** *Assume $L(E) > c_0 > 0$ for all $E$.*

*Let $E : I \to \mathbf{R}$ be a continuous function on a subinterval $I \subset [0, 1], |I| < 10^{-3}$, and $N$ a large integer and $0 < \kappa < 1$ satisfying the following conditions:*

$$\log\log \frac{1}{|I|} \ll \log N \tag{12.16}$$

$$\log\log \frac{N}{|I|} \ll \log\log \frac{1}{\kappa} \tag{12.16'}$$

*For each $x \in I$, there is a vector $\xi \in [e_j| |j| < N], \|\xi\| = 1$ s.t.*

$$\|(H(x) - E(x))\xi\| < \kappa \tag{12.17}$$

*Then*

$$\operatorname{mes} E(I) > e^{-(\log \frac{N}{|I|})^C} \tag{12.18}$$

**Proof.** The set $E(I)$ is an interval $[E_0 - \varepsilon, E_0 + \varepsilon]$, and our aim is to obtain a lower bound on $\varepsilon$. Fix

$$N \gg K = \left(\log \frac{N}{|I|}\right)^A \tag{12.19}$$

with $A$ a sufficiently large constant.

Since $L(E_0) > c_0$, we may apply the Green's function estimate from Proposition 7.19 at scale $K$. Thus, except for $x$ in a set of measure $< e^{-K^\sigma}$ ($\sigma > 0$ some constant), one of the intervals $\Lambda = \Lambda(x) = [1, K], [1, K-1], [2, K], [2, K-1]$ will satisfy

$$|G_\Lambda(E_0, x)(n_1, n_2)| < e^{-c|n_1 - n_2| + K^{1-}} \text{ for } n_1, n_2 \in \Lambda \tag{12.20}$$

Paving an interval $[-N, N] \subset \Lambda_1 \subset [-N-1, N+1]$ with size $K$ intervals $\Lambda$ satisfying (12.20) and applying the resolvent identity (see Lemma 10.33) gives the following:

For all $x$ outside a set $\Omega$ with

$$\text{mes}\,\Omega < Ne^{-K^\sigma}$$

there is an interval $[-N, N] \subset \Lambda_1 = \Lambda_1(x) \subset [-N-1, N+1]$ satisfying in particular

$$\|G_{\Lambda_1}(E_0, x)\| < e^K \tag{12.21}$$

Recalling (12.19), we may ensure that $\text{mes}\,\Omega < \text{mes}\,I$. Take $x \in I \backslash \Omega$ and $\xi$ the corresponding vector satisfying (12.17). Since $\xi \in [e_j | |j| < N]$ and choice of $\Lambda_1$

$$(H_{\Lambda_1} - E_0)\xi = (H - E_0)\xi = (H - E(x))\xi + 0(\varepsilon)$$

$$\|(H_{\Lambda_1}(x) - E_0)\xi\| < \kappa + \varepsilon$$

and from (12.21),

$$e^{-K} < \kappa + \varepsilon$$

Recalling (12.16') and (12.19), $e^{-K} \gg \kappa$, so that $\varepsilon > \frac{1}{2}e^{-\kappa}$. This proves (12.18).

**Remark.** Lemma 12.15 may be formulated simply on the basis of Green's function assumptions, independently of the Lyapounov exponent.

**Lemma 12.22.** Let $I \subset [0, 1]$ be an interval and $E(x) \in \text{Spec}\,H_N(x)$ a continuous function on $I$. Assume again that for all $x \in I$ there is $\xi \in [e_j | |j| < N]$, $\|\xi\| = 1$. s.t.

$$\|(H(x) - E(x))\xi\| < e^{-N^c} \tag{12.23}$$

(where $c > 0$ is some constant).

Let

$$\log N_1 \gg \log N \gg \log\log N_1$$

Then there is a system $(I', E_{I'})_{I' \in \mathcal{J}'}$ where $\mathcal{J}'$ is a collection of at most $N_1^C$ intervals $I' \subset I$ satisfying the previous assumptions with $N$ replaced by $N_1$, and moreover,

$$\text{mes}\left(\bigcup E_{I'}(I')\right) > \text{mes}\,E(I) - \frac{1}{N_1} \tag{12.24}$$

**Proof.** We assume $v$, the potential, a trigonometric polynomial to avoid some approximation arguments. Let $(\lambda_s = \lambda_s(x))_{|s| \leq N_1}$ be a continuous parametrization of $\text{Spec}\,H_{N_1}(x)$, and denote $\varphi_s = \varphi_s(x)$ the corresponding eigenfunctions (no regularity properties in $x$ will be involved). Fix $x \in I$. If $\xi$ is the vector from (12.24), write

$$\xi = \sum \langle \xi, \varphi_s \rangle \varphi_s$$

and

$$\begin{aligned}
H_N\xi = H_{N_1}\xi &= \sum \lambda_s \langle \xi, \varphi_s \rangle \varphi_s \\
&= E\xi + 0(e^{-N^c}) \\
&= E\sum \langle \xi, \varphi_s \rangle \varphi_s + 0(e^{-N^c})
\end{aligned}$$

Thus

$$\left(\sum |\lambda_s - E|^2 \langle \xi, \varphi_s \rangle^2\right)^{1/2} < e^{-N^c}$$

It follows that there is some $|s| \le N_1$ s.t.

$$|\langle \varphi_s, \xi \rangle| > \frac{1}{\sqrt{N_1}} \qquad (12.25)$$

and

$$|\lambda_s - E| < \sqrt{N_1}\, e^{-N^c} \qquad (12.26)$$

As in the proof of Theorem 10.1 (see (10.12)), we also may find some $\frac{N_1}{2} < j < N_1$ s.t.

$$|\langle \varphi_s, e_j \rangle| + |\langle \varphi_s, e_{j-1} \rangle| + |\langle \varphi_s, e_{-j} \rangle| + |\langle \varphi_s, e_{-j+1} \rangle| < e^{-2N_1^c}$$

(most $j$ do satisfy this in fact).

Since $\|P_N \varphi_s\| > \frac{1}{\sqrt{N_1}}$ by (12.25), the preceding implies that

$$\min_{\substack{\eta \in [e_k \mid |k| < j] \\ \|\eta\| = 1}} \|(H(x) - \lambda_s(x))\eta\| < N_1^{1/2} e^{-2N_1^c} < e^{-N_1^c}$$

or equivalently

$$\|[P_j (H(x) - \lambda_s(x))^* (H(x) - \lambda_s(x)) P_j]^{-1}\| > e^{2N_1^c} \qquad (12.27)$$

(the vector $\eta$ above is simply obtained as $P_j \varphi_s$).

Define for $|s| \le N_1$ and $\frac{N_1}{2} < j < N_1$ the set $\Gamma_{s,j} \subset I$ of $x$ for which (12.26) and (12.27) hold. Thus

$$I = \bigcup_{s,j} \Gamma_{s,j}$$

Observe (since $v$ was assumed to be a trigonometric polynomial) that $\zeta = \lambda_s$ (resp. $\lambda_s - E$) are continuous functions on $I$ satisfying an equation of the form

$$\zeta^d + \sum_{r<d} c_r(x) \zeta^r = 0 \qquad (12.28)$$

with $d = 2N_1 + 1$ (resp. $(2N + 1)(2N_1 + 1)$) and where the $c_r(x)$ are trigonometric polynomials of degree at most $N_1^C$. Hence the set (12.26) has at most $N_1^C$ components. Expressing (12.27) by Cramer's formula, a polynomial condition

$$P(x, \zeta) > 0 \qquad (12.29)$$

is obtained in $(x, \zeta = \lambda_s(x))$, where $\zeta$ satisfies (12.28). Therefore, condition (12.27) also leads to at most $N_1^C$ components. We may thus take each $\Gamma_{sj}$ as a union of at most $N_1^C$ intervals $I'$. For $I' \subset \Gamma_{sj}$, let $E_{I'} = \lambda_s$, for which by construction

$$\inf_{\substack{\eta \in [e_k \mid |k| < N] \\ \|\eta\| = 1}} \|(H(x) - E_{I'}(x))\eta\| < e^{-N_1^C}$$

for $x \in I'$. Let $\mathcal{J}'$ be the collection of all these $I'$-intervals; hence $\#\mathcal{J}' < N_1^C$. Clearly,

$$E(I) = \bigcup_{s,j} E(\Gamma_{s,j})$$

$$\subset \bigcup_{s,j} \bigcup_{I' \subset \Gamma_{s,j}} \lambda_s(I') + [-e^{-\frac{1}{2}N^c}, e^{-\frac{1}{2}N^c}]$$

$$= \bigcup_{I' \in \mathcal{J}'} E_{I'}(I') + [-e^{-\frac{1}{2}N^c}, e^{-\frac{1}{2}N^c}]$$

and hence

$$\mathrm{mes}\, E(I) \leq \mathrm{mes}\left(\bigcup E_{I'}(I')\right) + 4(\#\mathcal{J}')e^{-\frac{1}{2}N^c}$$

$$< \mathrm{mes}\left(\bigcup E_{I'}(I')\right) + e^{-\frac{1}{3}N^c}$$

This proves (12.24).

**Proof of Proposition 12.14.** Start from a large integer $N_0$, and consider the restriction $H_{N_0}$ of $H$ to $[-N_0, N_0]$. Let $\lambda_s = \lambda_s(x)$ be a continuous eigenvalue parametrization for $H_{N_0}(x)$. The same considerations as in the proof of Lemma 12.22 applied to $\xi = e_0$ permit us to obtain an interval $I_0 \subset [0,1]$ and a continuous function $E_0(x) \in \mathrm{Spec}\, H_{N_0}(x)$ ($E_0 = \lambda_s|_{I_0}$ for some $s$) satisfying

$$|I_0| > N_0^{-C}$$

and

$$\min_{\substack{\xi \in [e_k| |k| < N_0] \\ \|\xi\| = 1}} \|[H(x) - E_0(x)]\xi\| < e^{-N_0^c} \text{ for } x \in I_0$$

From Lemma 12.15,

$$\mathrm{mes}\, E_0(I_0) > e^{-(\log N_0)^C}$$

Starting from $(I_0, E_0)$, apply Lemma 12.22 with

$$\log\log N_0 \ll \log\log N_1 \ll \log N_0$$

to get a system $(I, E_I)_{I \in \mathcal{J}_1}$ satisfying the properties of the lemma. Next, repeat to each $I \in \mathcal{J}_1$ to obtain $(I, E_I)_{I \in \mathcal{J}_2}$ etc. Thus

$$\mathrm{mes}\left(\bigcup_{I \in \mathcal{J}_s} E_I(I)\right)$$

$$> \mathrm{mes}\left(\bigcup_{I \in \mathcal{J}_{s-1}} E_I(I)\right) - \tfrac{1}{N_s} > \cdots > \mathrm{mes}\, E_0(I_0) - \tfrac{1}{N_1} - \cdots - \tfrac{1}{N_s}$$

$$> \tfrac{1}{2}\mathrm{mes}\, E_0(I_0)$$

The property (12.23) implies that

$$\mathrm{dist}\left(E_I(x), \mathrm{Spec}\, H\right) < e^{-N_s^c}$$

if $x \in I \in \mathcal{J}_s$. Hence

$$\bigcap_s \bigcup_{I \in \mathcal{J}_s} E_I(I) \subset \mathrm{Spec}\, H$$

proving the result.

**Remarks.**

**1.** It follows from Proposition 12.14 and the results on Lyapounov exponents that if $H_\lambda(x) = \lambda v(x + n\omega)\delta_{nn'} + \Delta$, and $|\lambda| > \lambda_0$, then mes Spec $H_\lambda > 0$. The proof also enables us to establish this fact more generally for lattice Schrödinger operators

$$H_\lambda(x) = \lambda v(x + n\omega)\delta_{nn'} + S_\phi$$

$v$ real analytic on **T** and where $\Delta$ is replaced by a Toeplitz operator $S_\phi$ with real analytic symbol $\phi$ on **T**. (In Lemma 12.15 we apply directly the Green's function result Proposition 11.4–valid in this generality–without involving Lyapounov exponents).

**2.** Return to the Aubry duality described in Chapter 11. Thus, starting from

$$H_\lambda(x) = \lambda v(x + n\omega)\delta_{nn'} + \Delta \qquad (12.30)$$

(1-frequency case), we introduced

$$\widetilde{H}(x) = \cos(x + n\omega)\delta_{nn'} + \frac{\lambda}{2}S_v$$

Analyzing the construction, we see that

$$\text{Spec } \widetilde{H} = \frac{1}{2}\text{Spec } H$$

Since, by previous remark, mes Spec $\widetilde{H} > 0$ for $\lambda$ sufficiently small, it follows that

$$\text{mes } H_\lambda > 0$$

both for $|\lambda| > \lambda_0$ and for $|\lambda| < \lambda_1; \lambda_1, \lambda_0 > 0$.

**3.** In particular, in the Almost Mathieu case $H_\lambda(x) = \lambda \cos(x + n\omega) + \Delta$,

$$\text{mes Spec } H_\lambda > 0 \text{ for } |\lambda| \neq 2$$

It is known that in this particular case

$$\text{mes Spec } H_\lambda = 2\big|2 - |\lambda|\big| \qquad (12.31)$$

(see [L] for a discussion and references).

**4.** It is conjectured that for (12.30), $\lambda \neq 0$, $v$ nonconstant real analytic on **T** (1-frequency case), and $\omega DC$, Spec $H_\lambda$ has an empty interior.

This is known in the following cases. Let $v(x) = \cos x$. In the perturbative regime ($\lambda$ large or small), Spec $H_\lambda$ is a Cantor set [Sin]. If $\omega$ is an irrational Liouville number, rational approximations enable us to show that Spec $H_\lambda$ has no interior points (see [B-S]). If $\lambda = 2$, then mes Spec $H_\lambda = 0$ (see [G-J-L-S]), and the property is clear. In this situation, the renormalization group analysis from [H-S] enables a more precise description of the Cantor set for $\omega$'s with continued fraction expansion

$$\omega = \cfrac{1}{a_1 + \cfrac{1}{a_2 + \cfrac{1}{a_3 + \cdots}}}$$

satisfying

$$a_j > C \text{ for all } j \geq 1$$

where $C$ is a sufficiently large constant (see the remark in Chapter 8).

Very recently it was shown by J. Puig that the Almost Mathieu operator has Cantor spectrum for all $\lambda \neq 0$ and $\omega$ diophantine (preprint 2003).

# *References*

[B]  J. Bourgain. On the spectrum of lattice Schrödinger operators with deterministic potential, *J. Anal. Math.* 87 (2002), 37–75.

[L]  Y. Last. Almost everything about the Almost Mathieu operator, *Proc. XI, International Congress Math Phys., Paris 1994*, pp. 366–372. International Press, Cambridge, England, 1995.

[Sin]  Y. Sinai. Anderson localization for one-dimensional difference Schrödinger operator with quasi-periodic potential, *J. Stat. Phys.* 46 (1987), 861–909.

[B-S]  J. Bellissard, B. Simon. Cantor spectrum for the Almost Mathieu equation, *J. Funct. Anal.* 48(3) (1982), 408–419.

[G-J-L-S]  A. Gordon, S. Jitomirskaya, Y. Last, B. Simon. Duality and singular continuous spectrum in the Almost Mathieu equation, *Acta Math.* 178 (1997), 169–183.

[H-S]  B. Helffer, J. Sjöstrand. *Mémoires de la SMF* 34 (1988) and 39 (1989).

# Chapter Thirteen

## Point Spectrum in Multifrequency Models at Small Disorder

Consider Schrödinger operators on $\mathbf{Z}$

$$H_\lambda(x) = \lambda v(x + n\omega)\delta_{nn'} + \Delta \tag{13.1}$$

where $v$ is a nonconstant trigonometric polynomial on $\mathbf{T}^d$. We proved that if $\lambda > \lambda_0(v)$ then typically pure point spectrum with localization occurs. If $d = 1$, then for $\lambda < \lambda_1(v)$, one obtains purely (absolutely) continuous spectrum. These results are nonperturbative in the sense that $\lambda_0, \lambda_1$ do not depend on $\omega$ (which is always assumed diophantine).

It turns out that for $d \geq 2$, we cannot expect such nonperturbative statements for the continuous spectrum at small disorder. Notice that the dual model of (13.1) is a lattice Schrödinger operator on the $\mathbf{Z}^d$-lattice

$$\widetilde{H}_\lambda(x) = \cos(x + n.\omega) + \frac{\lambda}{2}S_v \tag{13.2}$$

Localization results for operators of the form (13.2) were obtained in [C-D] but are perturbative. Thus $\widetilde{H}_\lambda(x)$ satisfies Anderson localization (and $H_\lambda$ has only continuous spectrum) $x$ a.s. for $\lambda < \lambda_1(v, \omega)$. The following statement shows that the condition on $\lambda$ does depend on $\omega$ if $d \geq 2$.

**Theorem 13.3.** *Let $v = v(x_1, x_2)$ be a trigonometric polynomial on $\mathbf{T}^2$ with a nondegenerate local maximum. Then, for $\omega$ in a set of positive measure, $H = v(n\omega)\delta_{nn'} + \Delta$ has some point spectrum (with localized states) and $\mathrm{mes}(\sum_{p,p}) > 0$.*

Thus in the context of (13.1) $(d = 2)$, for any $\lambda \neq 0$, $\mathrm{mes}\left(\sum_{pp}(H_\lambda(x))\right) > 0$ for $(x, \omega) \in \mathbf{T}^4$ in a set of positive measure.

Notice that this does not also exclude the presence of continuous spectrum (in other energy ranges). In fact, one may produce examples with both spectral types $\left(\sum_{pp} \neq \phi, \sum_{ac} \neq \phi\right)$ in different energy regions (see [B2]). Also, if $\lambda$ is small, the phenomenon is nontypical in the sense that for $\sum_{pp}(H_\lambda)$ to be nonvoid, $\omega$ has to be restricted to a set of small (but positive) measure.

**Remarks.**
**1.** In Theorem 13.3, the frequency vector $\omega$ will be chosen small. Due to this fact, some care will be required when applying the methods developed earlier to produce localized states.
**2.** The point spectrum in Theorem 13.3 will appear at the edge of the spectrum.
**3.** Localization for $\mathbf{Z}^2$-operators has been established in [BGS] for the general

quasi-periodic class

$$H(x) = v(x_1 + n_1\omega_1, x_2 + n_2\omega_2) + \varepsilon\Delta$$

($\Delta$ = Laplacian on $\mathbf{Z}^2$), where $v$ is an arbitrary real analytic potential on $\mathbf{T}^2$ (such that the partial maps $v(x_1, \cdot)$ and $v(\cdot, x_2)$ are nonconstant). The method used is different from [C-D] but also perturbative; i.e., $\varepsilon < \varepsilon_0(v, \omega)$.

We will next summarize the main ideas in proving Theorem 13.3.

In order to avoid some inessential technical matters, consider the example

$$v(x_1, x_2) = \lambda(\cos x_1 + \cos x_2)$$

with $\lambda > 0$ small (the idea described below works in general). Thus

$$H_{\lambda,\omega}(x) = \lambda\big(\cos(x_1 + n\omega_1) + \cos(x_2 + n\omega_2)\big)\delta_{nn'} + \Delta \qquad (13.4)$$

Clearly, $\max \operatorname{Spec} H \leq 2(1+\lambda)$, and if $|\omega| = |\omega_1| + |\omega_2|$ is small, then $\max \operatorname{Spec} H \approx 2(1 + \lambda)$. First we will construct an energy interval $I$ near $2(1 + \lambda)$ and finitely many tiny subintervals $I_\alpha \subset I$ (not depending on $x$) s.t.

$$L(E) > c(\lambda) > 0 \text{ for all } E \in I \backslash \bigcup I_\alpha \qquad (13.5)$$

To be more precise about size, fixing a large integer $N_0$, we take

$$\log \frac{1}{|\omega|} \sim \log N_0 \qquad (13.6)$$

and

$$|I| > \rho(\lambda)$$

$$\log \log \frac{1}{|I_\alpha|} \sim \log N_0$$

$$\log(\#\{I_\alpha\}) \lesssim \log N_0$$

Next, we prove that

$$\operatorname{mes}\left(\sum(H) \cap I\right) > c_1(\lambda) > 0 \qquad (13.7)$$

so that necessarily (for $N_0$ large enough)

$$\operatorname{mes}\left(\sum(H) \cap (I \backslash \bigcup I_\alpha)\right) > 0 \qquad (13.8)$$

Also, invoking (13.5) and Proposition 12.3,

$$\sum\nolimits_{ac} (H(x)) \bigcap (I \backslash \bigcup I_\alpha)) = \phi \quad x \text{ a.s.}$$

and hence

$$\operatorname{mes}\left(\sum\nolimits_{pp} (H(x)) \bigcup \sum\nolimits_{sc} (H(x))\right) > 0 \quad x \text{ a.s.}$$

To get point spectrum, put $x = 0$ and restrict $\omega$ according to Remark (3) following Corollary 10.34. This ensures in particular that any extended state of $H_{\lambda,\omega}(0)$ with energy $E \in I \backslash \bigcup_\alpha I_\alpha$ is localized (notice that the $I_\alpha$ intervals do depend on $\omega$). Therefore,

$$(I \backslash \bigcup I_\alpha) \cap \sum\nolimits_c (H_\omega(0)) = \phi$$

and by (13.8),

$$\text{mes} \sum_{pp} \left(H_\omega(0)\right) > 0$$

Property (13.7) will be established using the constructive approach from Chapter 11. Consider the restriction $H_{N_0}$ of $H$ to $[-N_0, N_0]$. We will show that

$$\text{mes} \left( \bigcup_{|s|<\delta} \text{Spec} \, H_{N_0}\left(0, -\frac{\pi}{2} + s\right) \cap I \right) > \delta' \tag{13.9}$$

$(\delta, \delta' > 0$ depending on $\lambda)$ with eigenvectors $\xi \in [e_j| \, |j| \le N_0]$ that are exponentially small for $|j| > \frac{3}{4}N_0$. This, together with the Lyapounov exponent assumption (13.5), will permit us to prove (13.7) as Proposition 12.14 by iteration of Lemma 12.22.

As we mentioned, the fact that $\omega$ is small $\big($see (13.6)$\big)$ requires some care since the LDT for the fundamental matrix applies only at scales $N \gg N_0$.

Next, we pass to details. Consider (13.4) with $0 < \lambda < 1$ fixed (possibly small). Fix a large integer $N_0$, and let

$$\frac{1}{10^2 N_0} < \omega_1 < \frac{2}{10^2 N_0} \tag{13.10}$$

$$\frac{1}{N_0^2} < \omega_2 < \frac{2}{N_0^2} \tag{13.11}$$

Consider the matrix

$$A(s) = R_{[-N_0, N_0]} H\left(0, -\frac{\pi}{2} + s\right) R_{[-N_0, N_0]}$$
$$= \left(\lambda(\cos n\omega_1 + \sin(n\omega_2 + s))\delta_{nn'} + \Delta\right)_{|n|,|n'|\le N_0}$$

with $s \in [0, \delta]$.

Considering the vector $\zeta = \sum_{0<n<\sqrt{N_0}} e_n$, (13.10) and (13.11) imply

$$\langle A(0)\zeta, \zeta \rangle = \lambda \sum_{0<n<\sqrt{N_0}} (\cos n\omega_1 + \sin n\omega_2) + 2(\sqrt{N_0} - 1) = (2+\lambda)|\zeta|^2 + 0(1)$$

and hence

$$\max \text{Spec} \, A(0) > 2 + (1 - 10^{-6})\lambda$$

Considering a real analytic eigenvalue parametrization $\lambda_\alpha(s), \alpha = 0, 1, \ldots$ for $A(s)$, assume

$$\lambda(0) = \lambda_0(0) > 2 + (1 - 10^{-6})\lambda$$

Since

$$\frac{dA(s)}{ds} = \lambda \cos(s + n\omega_2)\delta_{nn'} = \lambda \cos s. Id + 0\left(\frac{1}{N_0}\right)$$

first-order eigenvalue variation implies that

$$\lambda_0'(s) > \frac{\lambda}{2} \text{ for } s \in [0, \delta]$$

Hence

$$\lambda_0([0, \delta]) = I$$

where $I$ is an interval satisfying

$$|I| \sim \delta\lambda \quad \text{and} \quad \min_{E \in I} E > 2 + (1 - 10^{-6})\lambda$$

Denote $\xi, \|\xi\| = 1$ an eigenvector of $A(s)$ with eigenvalue $E > 2 + (1 - 10^{-6})\lambda$. Our aim is to show that $|\xi_n|$ is small for $|n|$ near $N_0$. Observe that $A(s) - E$ has diagonal elements

$$d_n = \lambda\big( \cos n\omega_1 + \sin(n\omega_2 + s) \big) - E$$

satisfying, since $|n\omega_1| < \frac{1}{50}$,

$$\begin{aligned}
|d_n| &> E - \lambda \cos n\omega_1 + 0(\lambda\delta) \\
&> 2 + \lambda - 10^{-6}\lambda - \lambda(1 - \tfrac{1}{4}|n\omega_1|^2) + 0(\lambda\delta) \\
&> 2 + \tfrac{\lambda}{4}|n\omega_1|^2 - 10^{-6}\lambda + 0(\lambda\delta)
\end{aligned}$$

Hence, for $|n| > \frac{1}{2}N_0$,

$$|d_n| > 2 + \frac{\lambda}{16.10^4} - 10^{-6}\lambda + 0(\delta\lambda) > 2 + 10^{-6}\lambda$$

and a simple Neumann inversion argument shows that

$$|\xi_n| < e^{-c\lambda N_0} \quad \text{for } |n| > \frac{3N_0}{4}$$

In particular, for $s \in [0, \delta]$

$$\min_{\substack{\xi \in [e_j] \mid |j| < N_0] \\ \|\xi\| = 1}} \left\| \left( H\left(0, s - \frac{\pi}{2}\right) - \lambda_0(s) \right) \xi \right\| < e^{-c\lambda N_0} \tag{13.12}$$

Next, we turn our attention to the Lyapounov exponent $L(E)$ of (13.4). Our aim is to extract from $I$ at most $N_0^C$ subintervals $I_\alpha, |I_\alpha| < e^{-N_0^C}$, such that (13.5) holds on $I \backslash \bigcup I_\alpha$. Once this fact is established, we may find a subinterval $[a, b] \subset [0, \delta], b - a > N_0^{-C}$, such that for $s \in [a, b]$,

$$\lambda(s) \notin \bigcup I_\alpha$$

Hence

$$L\big(\lambda(s)\big) > c_0$$

Iteration of Lemma 12.22 enables one to prove that then

$$\text{mes}\,\big(\lambda([a, b]) \backslash \text{Spec}\, H\big) < \frac{1}{N_1} + \frac{1}{N_2} + \cdots \tag{13.13}$$

where

$$\log N_s \ll \log N_{s+1} \ll \log \log N_s$$

Thus

$$\text{mes}\,\big(\text{Spec}\, H \cap (I \backslash \bigcup I_\alpha)\big) > \frac{1}{2}\text{mes}\,\lambda([a, b]) > \frac{\lambda}{2}(b - a) > 0$$

which is (13.8). Properties (13.5) and (13.8) are thus obtained, which, as explained earlier, enables us to deduce Theorem 13.3.

An important point to notice is that when proving (13.13), use of the LDT to obtain the required Green's function estimates only appears at scales $N$ with $\log N \gtrsim \log N_1 \gg \log N_0$ and hence compatible with $\omega$ as in (13.10) and (13.11).

**Proof of (13.5).** Assume $E > 2 + (1 - 10^{-6})\lambda$. The diagonal of $H(x) - E$ is given by

$$d_n(x) = \lambda\big( \cos(x_1 + n\omega_1) + \cos(x_2 + n\omega_2)\big) - E$$

and hence, restricting $|n - \bar{n}| < N_0$,

$$|d_n(x)| > |E - \lambda\big( \cos(x_1 + \bar{n}\omega_1) + \cos(x_2 + \bar{n}\omega_2))| - \frac{\lambda}{10}$$

$$> 2 + \frac{\lambda}{2} \tag{13.14}$$

provided

$$\cos(x_1 + \bar{n}\omega_1) + \cos(x_2 + \bar{n}\omega_2) \leq 0$$

Denote $N_{00} = N_0^{10}$, say, and consider for fixed $x \in \mathbf{T}^2$ the orbit $x + n\omega, 1 \leq n \leq N_{00}$. From (13.10) and (13.11) it is clear that there is a collection of $N_0$-intervals $\Lambda_\alpha$ in $[1, N_{00}]$ for which (13.14) holds; thus

$$|G_{\Lambda_\alpha}(E, x)(n, n')| < \lambda^{-1} e^{-c\lambda|n-n'|} \text{ for } n, n' \in \Lambda_\alpha \tag{13.15}$$

and satisfying

$$\sum |\Lambda_\alpha \cap J| > \frac{1}{10}|J| \tag{13.16}$$

whenever $J \subset [1, N_{00}]$ is an interval of size $|J| > 100N_0$.

Moreover, assume that $E$ satisfies

$$\text{dist}\,\big(E, \text{Spec}\, H_\Lambda(x)\big) > e^{-N_0^{1/2}} \tag{13.17}$$

and hence

$$\|G_\Lambda(E, x)\| < e^{N_0^{1/2}}$$

for any subinterval $\Lambda \subset [1, N_{00}], |\Lambda| > N_0$.

It may then be shown using the resolvent identity (as in Lemma 10.33) that

$$|G_{[0,N_{00}]}(E, x)(n, n')| < e^{10N_0}$$

and

$$|G_{[0,N_{00}]}(E, x)(n, n')| < e^{-c(\lambda)|n-n'|} \text{ for } n, n' \in [0, N_{00}], |n - n'| > \frac{1}{10}N_{00}$$

(see Lemma 13.23 below).

Consequently,

$$|\det[H_{N_{00}}(x) - E]|^{-1} = |G_{[0,N_{00}]}(E, x)(0, N_{00})| < e^{-c(\lambda)N_{00}}$$

and thus

$$\frac{1}{N_{00}} \log \|M_{N_{00}}(x, E)\| > c(\lambda) \tag{13.18}$$

From condition (13.17), it is clear that (13.18) will hold for all $(x, E) \in \mathbf{T}^2 \times I$ except for a set of measure $< N_{00}^3 e^{-N_0^{1/2}}$. A Fubini argument therefore gives a subset $\mathcal{E}$ of $I$ such that

$$\operatorname{mes} \mathcal{E} < e^{-\frac{1}{2} N_0^{1/2}}$$

and if $E \notin \mathcal{E}$, then (13.18) holds for all $x$ in a set of measure $> \frac{1}{2}$.

Consequently,

$$L_{N_{00}}(E) > \frac{1}{2} c(\lambda) \text{ for } E \in I \backslash \mathcal{E}$$

Starting from scale $N_{00}$, the proof of Proposition 7.2 enables us to show that

$$|L(E) - L_{N_{00}}(E)| < N_{00}^{-1/2} \tag{13.19}$$

Hence

$$L(E) > \frac{1}{3} c(\lambda) \text{ for } E \in I \backslash \mathcal{E} \tag{13.20}$$

Notice again that at scales $N > N_{00} = N_0^{10}$, the LDT for $\frac{1}{N} \log \|M_N(x, E)\|$ applies.

It remains to obtain $\mathcal{E}$ as a union of at most $N_{00}^C = N_0^{10C}$ intervals. This is a consequence of the semialgebraic nature of the condition. One may define $\mathcal{E}$ as the set of $E \in I$ such that

$$\operatorname{mes} \left[ x \in [0, 1]^2 \mid \|M_{N_{00}}(x, E)\| < e^{c(\lambda) N_{00}} \right] > e^{-N_0^{1/4}} \tag{13.21}$$

The set

$$\mathcal{A} = \left[ (x, E) \in [0, 1]^2 \times I \mid \|M_{N_{00}}(x, E)\|_{HS}^2 < e^{2c(\lambda) N_{00}} \right]$$

is clearly semialgebraic of degree $< C N_{00}$.

If (13.21) holds, there is some $y \in [0, 1]^2$ such that $(x, E) \in \mathcal{A}$ for all $x$ with $|x - y| < e^{-10 N_0^{1/4}}$. Consider the set $\mathcal{B} \subset [0, 1]^2 \times [0, 1]^2 \times I$ of elements $(x, y, E)$ for which

$$\begin{cases} |x - y| < e^{-10 N_0^{1/4}} \\ (x, E) \notin \mathcal{A} \end{cases}$$

which is semialgebraic of degree $< C N_{00}$.

Define

$$\mathcal{E}_1 = \operatorname{Proj}_E C(\operatorname{Proj}_{y, E} \mathcal{B}) \tag{13.22}$$

It follows from the preceding that $\mathcal{E} \subset \mathcal{E}_1$ and, also for $E \in \mathcal{E}_1$

$$\operatorname{mes} \left[ x \in [0, 1]^2 \mid \|M_{N_{00}}(x, E)\|_{HS} < e^{c(\lambda) N_{00}} \right] > e^{-10 N_0^{1/4}}$$

Thus

$$\operatorname{mes} \mathcal{E}_1 < N_{00}^3 e^{-N_0^{1/2}} e^{10 N_0^{1/4}} < e^{-\frac{1}{2} N_0^{1/2}}$$

and $\mathcal{E}_1$ is a union of at most $N_{00}^C$ intervals.

This proves (13.5).

**Lemma 13.23.** *Let $A = v_n \delta_{nn'} + \Delta$ be an $N \times N$ matrix such that $|v_n| < C$. Assume there is a disjoint collection $\{I_\alpha\}$ of size $M$ intervals in $[1, N]$ $(M \ll N)$ s.t.*

$$\sum |I_\alpha| \geq cN$$

*and for each $\alpha$*

$$\|(R_{I_\alpha} A R_{I_\alpha})^{-1}\| < e^{M^{1-}} \tag{13.24}$$

*and*

$$|(R_{I_\alpha} A R_{I_\alpha})^{-1}(n, n')| < e^{-c|n-n'|} \text{ for } |n - n'| > \frac{M}{10} \tag{13.25}$$

*Assume further that for any interval $I \subset [1, N]$*

$$\|(R_I A R_I)^{-1}\| < e^{M^{1/2}} \tag{13.26}$$

*holds. Then*

$$|A^{-1}(n, n')| < e^{10M - c'd(n,n')} \tag{13.27}$$

*where for $n < n'$*

$$d(n, n') \equiv \sum_\alpha |[n, n'] \cap I_\alpha| \tag{13.28}$$

**Proof.** It is again an application of the resolvent identity.

Denote $J = [1, N]$. Assume that $k_1 < k_2 \in J$ and

$$d(k_1, k_2) > 10M$$

**Case 1.** $k_1 \in I_\alpha$ for some $\alpha$ and

$$\text{dist}\,(k_1, \partial I_\alpha) > \frac{M}{5}$$

It follows from the resolvent identity applied to the decomposition $J = I_\alpha \cup (J \backslash I_\alpha)$ that

$$
\begin{aligned}
|A^{-1}(k_1, k_2)| &\leq \sum_{\substack{k_3 \in \partial I_\alpha, k_4 \in J \backslash I_\alpha \\ |k_3 - k_4| = 1}} |(R_{I_\alpha} A R_{I_\alpha})^{-1}(k_1, k_3)|\, |A^{-1}(k_4, k_2)| \\
&\leq \sum_{\substack{k_3 \in \partial I_\alpha, k_4 \in J \backslash I_\alpha \\ |k_3 - k_4| = 1}} e^{-c|k_1 - k_3|}\, |A^{-1}(k_4, k_2)| \\
&\leq \max_{\frac{M}{5} \leq |k_1 - k'| \leq M} e^{-\frac{c}{2}|k_1 - k'|}\, |A^{-1}(k', k_2)| \\
&\leq e^{-\frac{c}{10}M} \max_{|k_1 - k'| \leq M} |A^{-1}(k', k_2)| \tag{13.29}
\end{aligned}
$$

by (13.25).

**Case 2.** Assume $I_\alpha = [a_\alpha, a_\alpha + M]$ and $I_\beta = [a_\beta, a_\beta + M]$ consecutive intervals s.t.

$$k_1 \in \left[a_\alpha + M - \frac{M}{5}, a_\beta + \frac{M}{5}\right]$$

Put

$$I = \left[ a_\alpha + M - \frac{M}{4}, a_\beta + \frac{M}{4} \right]$$

Apply the resolvent identity to the decomposition $J = I \cup (J \backslash I)$ to obtain

$$|A^{-1}(k_1, k_2)| \leq \sum_{\substack{k_3 \in I, k_4 \in J \backslash I \\ |k_3 - k_4| = 1}} |(R_I A R_I)^{-1}(k_1, k_3)| \, |A^{-1}(k_4, k_2)|$$

$$\leq e^{2M^{1/2}} \max_{k_4 \notin I, \text{dist } (k_4, I) = 1} |A^{-1}(k_4, k_2)| \qquad (13.30)$$

Clearly, $k_4 \in I_\alpha \cup I_\beta$, and if $k_4 \in I_\beta$, say, then dist $(k_4, \partial I_\beta) > \frac{M}{5}$. Thus, from the Case 1 estimate (13.29),

$$|A^{-1}(k_4, k_2)| \leq e^{-\frac{c}{10}M} \max_{|k_4 - k'| \leq M} |A^{-1}(k', k_2)|$$

and substitution in (13.30) yields

$$|A^{-1}(k_1, k_2)| < e^{-\frac{c}{11}M} \max_{d(k', k_2) > d(k_1, k_2) - 2M} |A^{-1}(k', k_2)| \qquad (13.31)$$

We use here the fact that

$$d(k_1, k_2) \leq d(k_1, k_4) + |k_4 - k'| + d(k', k_2)$$
$$\leq \frac{M}{5} + d(a_\alpha + M, a_\beta) + \frac{M}{4} + 1 + M + d(k', k_2)$$

and $d(a_\alpha + M, a_\beta) = 0$, by definition (13.28). Thus estimate (13.31) holds in both Cases 1 and 2. Straightforward iteration then permits us to establish (13.27) for some constant $c'$.

**Remark.** In [B2], examples are produced of the form

$$H_{\lambda_1, \lambda_2, \omega} = (\lambda_1 \cos n\omega_1 + \lambda_2 \cos n\omega_2) \delta_{nn'} + \Delta$$

such that for $\omega = (\omega_1, \omega_2)$ in a set of (small) positive measure, $\sum_{ac}(H) \neq \phi$ and mes $\left( \sum_{pp}(H) \right) > 0$. Again here the frequency vector $\omega$ is chosen to be small.

See [F-K] for results on quasi-periodic Schrödinger operators on $\mathbf{R}$ in this respect.

# *References*

[B1, 2]  J. Bourgain. On the spectrum of lattice Schrödinger operators with deterministic potential, I, II, *J. Analyse Math.* 87 (2002), 37–75, and 88 (2002),
221–254.

[BGS]  J. Bourgain, M. Goldstein, W. Schlag. Anderson localization for Schrödinger
operators on $\mathbf{Z}^2$ with quasi-periodic potential, *Acta Math.* 188 (2002), 41–86.

[C-Di]  V. Chulaevsky, E. Dinaburg.  Methods of KAM theory for long-range
quasi-periodic operators on $\mathbf{Z}^n$. Pure point spectrum, *Comm. Math. Physics*
153(3) (1993), 559–557.

[F-K]  A. Fedorov, F. Klopp.  Transition d'Anderson pour des opérateurs de
Schrödinger quasiperiodiques en dimension 1, *Séminaire: Equations aux
Dérivées Partielles*, 1998–1999, Expo. IV.

# Chapter Fourteen

## A Matrix-Valued Cartan-Type Theorem

The main result of this chapter is as follows:

**Proposition 14.1.** *Let $A(\sigma)$ be a self-adjoint $N \times N$ matrix function of a real parameter $\sigma \in [-\delta, \delta]$, satisfying the following conditions*

(i) *$A(\sigma)$ is real analytic in $\sigma$, and there is a holomorphic extension to a strip*

$$|\operatorname{Re} z| < \delta, |\operatorname{Im} z| < \gamma \qquad (14.2)$$

    *satisfying*

$$\|A(z)\| < B_1 \qquad (14.3)$$

(ii) *For each $\sigma \in [-\delta, \delta]$, there is a subset $\Lambda \subset [1, N]$ s.t.*

$$|\Lambda| < M \qquad (14.4)$$

    *and*

$$\|(R_{[1,N]\backslash\Lambda} A(\sigma) R_{[1,N]\backslash\Lambda})^{-1}\| < B_2 \qquad (14.5)$$

(iii)

$$\operatorname{mes}\left[\sigma \in [-\delta, \delta] \big| \, \|A(\sigma)^{-1}\| > B_3\right] < 10^{-3}\gamma(1+B_1)^{-1}(1+B_2)^{-1} \qquad (14.6)$$

    *Then, letting*

$$\kappa < (1 + B_1 + B_2)^{-10M}$$

    *we have*

$$\operatorname{mes}\left[\sigma \in \left[-\frac{\delta}{2}, \frac{\delta}{2}\right] \Big| \, \|A(\sigma)^{-1}\| > \frac{1}{\kappa}\right] < e^{-\frac{c \log \kappa^{-1}}{M. \log(M+B_1+B_2+B_3)}} \qquad (14.7)$$

**Proof.** Denote

$$\delta_1 = 10^{-2}\gamma(1 + B_1)^{-1}(1 + B_2)^{-1}$$

Fix $\sigma_0 \in \left[-\frac{\delta}{2}, \frac{\delta}{2}\right]$. If $z \in \mathbf{C}, |z - \sigma_0| < \delta_1$, it follows that

$$\|A(z) - A(\sigma_0)\| \le 2B_1\gamma^{-1}\delta_1 < \frac{1}{50}(1 + B_2)^{-1}$$

and hence, if $\Lambda \subset [1, N]$ is the index set associated to $\sigma_0$, we obtain from (14.5) (and a standard Neumann series argument)

$$\|(R_{\Lambda^c} A(z) R_{\Lambda^c})^{-1}\| < 2B_2 \text{ for } |z - \sigma_0| < \delta_1 \qquad (14.8)$$

$(\Lambda^c = [1, N] \backslash \Lambda)$.

Denote for $|z - \sigma_0| < \delta_1$ the analytic matrix-valued function (with index set $\Lambda$)

$$B(z) = R_\Lambda A(z) R_\Lambda - R_\Lambda A(z) R_{\Lambda^c} (R_{\Lambda^c} A(z) R_{\Lambda^c})^{-1} R_{\Lambda^c} A(z) R_\Lambda \quad (14.9)$$

satisfying by (14.3) and (14.8)

$$\|B(z)\| < 3B_1^2 B_2 \quad (14.10)$$

If $B(\sigma)$ is invertible, so is $A(\sigma)$ and (see [BGS], Lemma 4.8)

$$\|B(\sigma)^{-1}\| \lesssim \|A(\sigma)^{-1}\| \lesssim (1 + \|(R_{\Lambda^c} A(\sigma) R_{\Lambda^c})^{-1}\|^2)(1 + \|B(\sigma)^{-1}\|)$$
$$\lesssim (1 + 10B_2^2)(1 + \|B(\sigma)^{-1}\|) \quad (14.11)$$

For $z = \sigma \in \mathbf{R}$, $B(\sigma)$ is self-adjoint, and hence

$$|\det B(\sigma)| = \prod_{\lambda \in \operatorname{Spec} B(\sigma)} |\lambda| \geq \|B(\sigma)^{-1}\|^{-M} \quad (14.12)$$

Also, by Cramer's rule,

$$\|B(\sigma)^{-1}\| \leq \frac{\|B(\sigma)\|^{|\Lambda|}}{|\det B(\sigma)|} \leq \frac{(CB_1^2 B_2)^M}{|\det B(\sigma)|} \quad (14.13)$$

Consider the subharmonic function in $z$, $|z| < 1$,

$$u(z) = \log |\det B(\sigma_0 + \delta_1 z)|$$

with, again by (14.10), upper bound

$$u(z) \leq \log(CB_1^2 B_2)^M \lesssim M \log(M + B_1 + B_2) \quad (14.14)$$

To obtain a lower bound at some point, invoke (14.6). From the assumption, there is clearly some $\sigma_1 \in [-\delta, \delta]$, $|\sigma_0 - \sigma_1| < \frac{\delta_1}{10}$, such that

$$\|A(\sigma_1)^{-1}\| \leq B_3$$

Hence, by (14.11),

$$\|B(\sigma_1)^{-1}\| \lesssim B_3$$

and from (14.12),

$$|\det B(\sigma_1)| > B_3^{-M}$$

Consequently, denoting $a = \frac{\sigma_1 - \sigma_0}{\delta_1}$, $|a| < \frac{1}{10}$,

$$u(a) > -M \log B_3 \quad (14.15)$$

It follows from Jensen's inequality that for $0 < \rho < \frac{1}{2}$,

$$\int u(a + \rho e^{i\theta}) d\theta \geq u(a)$$

and hence

$$\int_{|z - a| < \frac{1}{2}} u(z) > -M \log B_3$$

Taking also (14.14) into account, it follows that

$$\|u\|_{L^1[|z-a|<\frac{1}{2}]} \lesssim M \log(M + B_1 + B_2 + B_3)$$

From the Riesz-representation theorem,

$$\frac{1}{M \log(M + B_1 + B_2 + B_3)} u(z) = \int \log|z - w| \mu(dw) + 0(1) \text{ for } |z - a| < \frac{1}{4}$$

with $\mu \in M_+(|z - a| < \frac{1}{2})$, $\|\mu\| < C$, and in particular,

$$\|u\|_{BMO(a-\frac{1}{4},a+\frac{1}{4})} \lesssim M \log(M + B_1 + B_2 + B_3)$$

John-Nirenberg's estimate implies that

$$\text{mes} \left[\sigma \big| |\sigma - \sigma_1| < \tfrac{1}{4}\delta_1 \text{ and } |\det B(\sigma)| < \kappa\right]$$
$$= \delta_1 \text{mes} \left[x \big| |x - a| < \tfrac{1}{4} \text{ and } u(x) < -\log \tfrac{1}{\kappa}\right]$$
$$< \delta_1 \exp\left[-\frac{c \log \kappa^{-1}}{M \log(M + B_1 + B_2 + B_3)}\right]$$

Hence

$$\text{mes} \left[\sigma \big| |\sigma - \sigma_0| < \frac{1}{8}\delta_1 \text{ and } |\det B(\sigma)| < \kappa\right]$$
$$< \delta_1 \exp\left[-c\frac{\log \kappa^{-1}}{M \log(M + B_1 + B_2 + B_3)}\right] \tag{14.16}$$

If $|\sigma - \sigma_0| < \frac{\delta_1}{8}$, $|\det B(\sigma)| \geq \kappa$, (14.11) and (14.13) imply

$$\|A(\sigma)^{-1}\| \lesssim (1 + B_2^2)(1 + \kappa^{-1}(CB_1^2 B_2)^M) < \kappa^{-2}$$

Since $\sigma_0 \in \left[-\frac{\delta}{2}, \frac{\delta}{2}\right]$ was

$$\text{mes} \left[\sigma \in \left[-\frac{\delta}{2}, \frac{\delta}{2}\right] \big| \|A(\sigma)^{-1}\| > \kappa^{-2}\right] < \exp\left[-c\frac{\log \kappa^{-1}}{M \log(M + B_1 + B_2 + B_3)}\right]$$

which proves Proposition 14.1.

**Remarks.**
**1.** In application, $\log \frac{1}{\kappa} = o(N)$, and to obtain a nontrivial estimate from (14.7), we thus need to assume that

$$M < N^\rho \quad B_1, B_2, B_3 < e^{N^{\rho'}}$$

where $\rho, \rho' \geq 0$ satisfy $\rho + \rho' < 1$.
**2.** In the previous argument we could have alternatively invoked Cartan's result directly, once (14.14) and (14.15) established the (scalar) subharmonic function $u$.
**3.** Proposition 14.1 easily generalizes to matrix-valued functions $A(\sigma)$ depending real analytically on a multiparameter $\sigma$. For instance, if $\sigma = (\sigma_1, \sigma_2)$, estimate (14.7) should be replaced by

$$\text{mes} \left[\sigma \big| \|A(\sigma)^{-1}\| > \kappa^{-1}\right] < \exp\left[-c\left(\frac{\log \kappa^{-1}}{M \log(M + B_1 + B_2 + B_3)}\right)^{1/2}\right] \tag{14.17}$$

**4.** Some comments on conditions (ii) and (iii) in Proposition 14.1.

We always assume $A(\sigma)$ with exponential off-diagonal decay

$$|A(\sigma)(n, n')| < e^{-c_0|n-n'|} \tag{14.18}$$

Let

$$N_0 \ll N_1 \ll N$$

be 2 scales. For given $\sigma$, define $\mathcal{J}(\sigma)$ as the collection of size-$N_0$ intervals $I \subset [1, N]$ s.t., denoting $A_I = R_I A R_I$

$$\|A_I(\sigma)^{-1}\| < e^{N_0^{1-}} \tag{14.19}$$

and

$$|A_I(\sigma)^{-1}(n, n')| < e^{-c|n-n'|} \text{ for } n, n' \in I, |n - n'| > \frac{N_0}{10} \tag{14.20}$$

Assume that there are at most $M < \frac{N}{N_1}$ intervals $I, |I| = N_0$ s.t. $I$ is not in $\mathcal{J}(\sigma)$. Let $\{I_\alpha\}$ be a partition of $[1, N]$ in $N_0$-intervals and $\Lambda$ the union of the $I_\alpha$ for which there is an $N_0$-interval $I \subset [1, N]$, $I \cap I_\alpha \neq \phi$ and $I \notin \mathcal{J}(\sigma)$. Thus (14.4) holds, and from (14.18) through (14.20) and the resolvent identity, (14.5) follows with $B_2 = e^{N_0}$.

To establish (iii), we proceed as follows. Assume that it is established that for any $N_1$-interval $J \subset [1, N]$ the properties

$$\|A_J(\sigma)^{-1}\| < e^{N_1^{1-}} \tag{14.21}$$

and

$$|A_J(\sigma)^{-1}(n, n')| < e^{-c|n-n'|} \text{ for } n, n' \in J, |n - n'| > \frac{1}{10}N_1 \tag{14.22}$$

hold, except for $\sigma$ in a set of measure less than $10^{-3}N^{-1}\gamma B_1^{-1}e^{-N_0}$. Covering then $[1, N]$ by size-$N_1$ intervals, another application of the resolvent identity implies (14.6) with $B_3 = e^{N_1}$.

In this setting, estimate (14.7) becomes

$$\text{mes}\left[\sigma \in \left[-\frac{\delta}{2}, \frac{\delta}{2}\right] \mid \|A(\sigma)^{-1}\| > e^K\right] < e^{-c\frac{K}{M.N_1}} \tag{14.23}$$

**5.** The conclusion (14.7) in Proposition 14.1 only involves a bound on the inverse $A(\sigma)^{-1}$, while a multiscale inductive argument also requires off-diagonal decay estimates, (see (14.19) through (14.22)). This is achieved as follows:

We assume (14.18). Fix $\sigma$, and let $\Lambda = \bigcup I_\alpha, |I_\alpha| = N_0$, the set introduced in the preceding remark. Assume that for a fixed $0 < \rho < 1$

$$|\Lambda| = M < N^\rho \tag{14.24}$$

and $N_0 = N^{\rho_0}, N_1 = N^{\rho_1}$ with

$$\rho_0, \rho_1 \leq \frac{1}{10}(1 - \rho)^2$$

Define

$$\bar{N} = N^{\frac{1-\rho}{4}}$$

If $J \subset [1, N]$ is an $\bar{N}$-interval and $J \cap \Lambda = \phi$, it follows from definition of $\Lambda$ and the resolvent identity that

$$\|A_J(\sigma)^{-1}\| < e^{N_0} \tag{14.25}$$

and

$$|A_J(\sigma)^{-1}(n, n')| < e^{-c|n-n'|} \text{ for } n, n' \in J, |n - n'| > \frac{1}{10}\bar{N} \tag{14.26}$$

Call such $\bar{N}$-interval "good." Clearly, there are at most $N^\rho$ disjoint bad intervals. Assume in addition to (14.24) that

$$|\Lambda \cap J'| < |J'|^\rho \tag{14.27}$$

also holds for any interval $J' \subset [1, N]$ of size $|J'| = L > \bar{N}$.

To each of these intervals $J'$, apply Proposition 14.1 with $N$ replaced by $L > N^{\frac{1-\rho}{4}}$. By (14.27), condition (14.4) holds with $M = L^\rho$ and (14.5) with $B_2 = e^{N_0}$. Condition (14.6) is established as before with $B_3 = e^{N_1}$. Estimate (14.23) then implies that

$$\text{mes}\left[\sigma \in \left[-\frac{\delta}{2}, \frac{\delta}{2}\right] \mid \|A_{J'}(\sigma)^{-1}\| > e^{L^{\frac{1+\rho}{2}}}\right] < e^{-c\frac{L^{\frac{1+\rho}{2}}}{L^\rho \cdot N^{\rho_1}}}$$

$$< e^{-cL^{\frac{1-\rho}{2} - \frac{2}{5}(1-\rho)}}$$

$$= e^{-cL^{\frac{1-\rho}{10}}}$$

Thus the bound

$$\|A_{J'}(\sigma)^{-1}\| < e^{|J'|^{\frac{1+\rho}{2}}} \tag{14.28}$$

may be obtained for any subinterval $J' \subset [1, N], |J'| > \bar{N}$, excluding a $\sigma$-set of measure at most

$$N^2 e^{-c\bar{N}^{\frac{1-\rho}{10}}} < e^{-N^{10^{-2}(1-\rho)^2}} \tag{14.29}$$

We then may apply the following lemma, which is a variant of Lemma 13.23, to conclude that for these $\sigma$,

$$|A(\sigma)^{-1}(n, n')| < e^{-(c-)|n-n'|} \text{ for } |n - n'| > \frac{N}{10} \tag{14.30}$$

**Lemma 14.31.** *Assume $A$ an $N \times N$ matrix satisfying*

$$|A_{n,n'}| < e^{-c_0|n-n'|} \tag{14.32}$$

*Let $\bar{N} = N^\tau$ for some $0 < \tau < 1$. Assume that for any interval $J' \subset [1, N], |J'| = L \geq \bar{N}$,*

$$\|A_{J'}^{-1}\| < e^{L^b} \tag{14.33}$$

*where $0 < b < 1, \tau + b < 1$. Call an $\bar{N}$-interval $J$ "good" if, moreover,*

$$|A_J^{-1}(n, n')| < e^{-c|n-n'|} \text{ for } n, n' \in J, |n - n'| > \frac{\bar{N}}{10} \tag{14.34}$$

*(take $0 < c < \frac{1}{10}c_0$). Assume that there are at most $N^b$ disjoint bad $\bar{N}$-intervals. Under these assumptions*

$$|A^{-1}(n, n')| < e^{-c'|n-n'|} \text{ for } |n - n'| > \frac{N}{10} \qquad (14.35)$$

where $c' = c - N^{-\kappa}, \kappa = \kappa(\tau, b) > 0$.

We then apply Lemma 14.31 with $A = A(\sigma)$, $b = \max(\rho, \frac{1+\rho}{2}) = \frac{1+\rho}{2}, \tau = \frac{1-\rho}{4}$. The size of the exceptional $\sigma$-set at scale $N$ is bounded by (14.29). With such estimate at scale $N_1$, satisfying (14.6) according to the discussion in Remark (4) will require us to take $N_0 < N_1^{10^{-3}(1-\rho)^2}$. Hence, for the scales $N_0, N_1$, one may take

$$N_0 = N^{10^{-3}a(1-\rho)^4} \text{ and } N_1 = N^{a(1-\rho)^2} \qquad (14.36)$$

with $0 < a < \frac{1}{10}$ a parameter. The main differences between Lemmas 13.23 and 14.31 is that the off-diagonal of $A$ is not given by $\Delta$ but assumed to satisfy a more general assumption (14.32) (this is a minor technical point). Also, some additional care is needed to preserve essentially the constant $c'$ in (14.35). See [B] for details.

**6.** The method described in this chapter does have applications to lattice Schrödinger operators on $\mathbf{Z}^D$, $D > 1$. See, for instance, [BSG] for results on perturbative quasi-periodic localization on the $\mathbf{Z}^2$-lattice obtained along these lines.

# References

[B]  J. Bourgain. Estimates on Green's functions, localization and the quantum kicked rotor model, *Annals of Math.* 156(1) (2002), 249–294.

[BGS]  J. Bourgain, M. Goldstein, W. Schlag. See Chapter 13.

# Chapter Fifteen

## Application to Jacobi Matrices Associated with Skew Shifts

We consider 1D lattice Schrödinger operators $H(x), x \in \mathbf{T}^d$ associated with a skew shift transformation $T : \mathbf{T}^d \to \mathbf{T}^d$, and thus $H_{m+1,n+1}(x) = H_{m,n}(Tx)$. To simplify matters, let $d = 2$ and

$$T : \mathbf{T}^2 \to \mathbf{T}^2 : (x_1, x_2) \mapsto (x_1 + x_2, x_2 + \omega)$$

(the method applies equally well to higher-dimensional skew shift extensions).

We always assume $\omega$ satisfying a DC. To avoid additional parameters, assume, say,

$$\|k.\omega\| > c|k|^{-2} \text{ for } k \in \mathbf{Z}\backslash\{0\} \tag{15.0}$$

$H(x)$ will be given by

$$H(x) = V(T^n x)\delta_{nn'} + \delta\Delta \tag{15.1}$$

where $V$ is a real nonconstant trigonometric polynomial on $\mathbf{T}^2$.

More generally, we will consider $H(x)$ of the form

$$H_{nn}(x) = V(T^n x) \tag{15.2}$$

and for $m \neq n$

$$H_{mn}(x) = \phi_{m-n}(T^m x) + \overline{\phi_{n-m}(T^n x)} \tag{15.3}$$

where again

$$V \text{ is a real nonconstant trigonometric polynomial on } \mathbf{T}^2 \tag{15.4}$$

$$\phi_k \text{ is a trigonometric polynomial of degree } < |k|^{C_1} \tag{15.5}$$

$$\|\phi_k\|_\infty < \delta\, e^{-|k|} \tag{15.6}$$

Our purpose is to obtain Green's function estimates and (dynamical) localization results. The model (15.1) also may be treated by the transfer matrix approach, (see [BGS]), but no nonperturbative results are known so far in the skew shift case.

The model (15.2) through (15.6) is of importance to our main application, which is the kicked rotor equation. This is the linear Schrödinger equation (1.4) with periodic time-dependent potential

$$i\frac{\partial u}{\partial t} + a\frac{\partial^2 u}{\partial x^2} + ib\frac{\partial u}{\partial x} + \kappa\left[\cos x \cdot \sum_{n \in \mathbf{Z}} \delta(t - n)\right]u = 0 \tag{15.7}$$

It's monodromy matrix $W$, as we will see, turns out to be of the preceding form (more precisely, $W + W^*$ satisfies the description (15.2) through (15.6)).

Obtaining the Green's function bounds will be an application of the preceding chapter (this approach works in a much more general context than the fundamental matrix technique, but results are perturbative). Proving localization and dynamical localization will then be achieved following the same scheme as in Chapter 10 (for the shift model) based on semialgebraic set theory. Of course, the required facts about intersections of semialgebraic sets and skew shift orbits will have to be established.

Our first goal is thus to prove

**Proposition 15.8.** *Let $H$ be given by (15.2) through (15.6) with $\delta$ in (15.6) suffi - ciently small. Then, for all suffi ciently large $N$ and energy $E$, we have that*

$$|G_N(E, x)| < e^{N^{1-}} \tag{15.9}$$

*and*

$$|G_N(E, x)(n, n')| < e^{-\frac{1}{100}|n-n'|} \text{ for } |n - n'| > \frac{N}{10} \tag{15.10}$$

*except for $x \in \Omega_N(E) \subset \mathbf{T}^2$ satisfying*

$$\operatorname{mes} \Omega_N(E) < e^{-N^\sigma} \tag{15.11}$$

*(for some $\sigma > 0$ that will be specifi ed later).*

The proof is multiscale and perturbative.

To start off, take $N_0$ a large integer. For $\delta = \delta(N_0)$ small enough, one then has that (for any $0 < \theta < 1$)

$$|G_{N_0}(E, x)(n, n')| < e^{N_0^\theta - \frac{1}{2}|n-n'|} \text{ for } n, n' \in [1, N_0] \tag{15.12}$$

except for $x$ in $\Omega_{N_0,\theta}(E)$ with $\operatorname{mes} \Omega_{N_0,\theta}(E) < e^{-cN_0^\theta}$.

Indeed, from assumption (15.4) and Lojasiewicz' inequality it follows that

$$\operatorname{mes} [x \in \mathbf{T}^2 | \, |V(x) - E| < \gamma] \lesssim \gamma^c$$

for all $\gamma > 0$ and where $c > 0$ is a constant depending on $V$ (not on $E$). Thus

$$\operatorname{mes} [x \in \mathbf{T}^2 | \min_{0 \leq n \leq N_0} |V(T^n x) - E| < \gamma] < N_0 \gamma^c$$

If $\min_{0 \leq n \leq N_0} |V(T^n x) - E| > \gamma > C\delta$, we get by Neumann expansion and (15.6)

$$|G_{[0,N_0]}(E, x)(m, n)| \leq$$

$$\gamma^{-1} \left[ 1 + \sum_{s \geq 1} \left( \frac{\delta}{\gamma} \right)^s \left( \sum_{1 \leq n_1, \ldots, n_{s-1} \leq N_0} e^{-(|m-n_1|+\cdots+|n_{s-1}-n|)} \right) \right]$$

$$< \gamma^{-1} e^{-\frac{1}{2}|m-n|}$$

Putting $\gamma < e^{-N_0^\theta}$ and $\delta = e^{-N_0}$, (15.12) follows.

The inductive step is achieved using the analysis from Chapter 14. The parameter $\sigma = (x_1, x_2) \in \mathbf{T}^2$ and $A(x) = H_N(x) - E$ has entries given by trigonometric polynomials of degree $< N^{1+C_1}$. This follows from (15.5) and the fact that

$$T^n x = (x_1 + nx_2 + \frac{n(n-1)}{2}\omega, x_2 + n\omega)$$

Thus, in conditions (14.2) and (14.3) on the analytic extension, take $\gamma = N^{-C_1-2}$, $B_1 = C$. Following Chapter 14, consider a scale $N_0$, $\log N_0 < \frac{1}{10} \log N$, $\log N_0 \sim \log N$ and to be specified later. Following Remarks (4) and (5) in Chapter 14, the issue is to establish (14.24) through (14.27) for some fixed, $\rho < 1$. Thus we need to show there is $0 < \rho < 1$ such that

$$|\Lambda \cap J| < |J|^\rho \tag{15.13}$$

holds for any subinterval $J \subset [1, N]$ of size $|J| = L > N^{\frac{1-\rho}{4}}$ and where $\Lambda$ is the complement of sites $n_0$ s.t., if $I = [n_0, n_0 + N_0]$

$$\|A_I(x)^{-1}\| < e^{N_0^{1-}} \tag{15.14}$$

$$|A_I(x)^{-1}(n, n')| < e^{-c|n-n'|} \text{ for } n, n' \in I, |n - n'| > \frac{N_0}{10} \tag{15.15}$$

$\big($condition (14.19) and (14.20)$\big)$. Here $x \in \mathbf{T}^2$ is fixed, and the preceding should hold uniformly. Also,

$$A_I(x) = H_{N_0}(T^{n_0}x) - E$$

Letting $\Omega_{N_0}(E)$ be the complement of the set of $x \in \mathbf{T}^2$ for which

$$\|G_{N_0}(E, x)\| < e^{N_0^{1-}} \tag{15.16}$$

and

$$|G_{N_0}(E, x)(n, n')| < e^{-\frac{1}{100}|n-n'|} \text{ for } |n - n'| > \frac{N_0}{10} \tag{15.17}$$

we need thus to estimate

$$\#\{n \in J \big| T^n x \in \Omega_{N_0}(E) \ (\text{mod } 1)\} \tag{15.18}$$

From (15.11) and the induction hypothesis, we have

$$\operatorname{mes} \Omega_{N_0}(E) < e^{-N_0^\sigma} \tag{15.19}$$

Expressing again $G_{N_0}(E, x)$ as a ratio of determinants, $\Omega_{N_0}(E)$ clearly may be viewed as a semialgebraic set of degree at most $N_0^{3+C_1}$.

Fix $\varepsilon > e^{-\frac{1}{10}N_0^\sigma}$ (in its later choice, $\log \frac{1}{\varepsilon} \sim \log N$, and this condition will obviously hold). It follows from the uniformization theorem for semialgebraic sets (Theorem 9.4) that $\partial \Omega_{N_0}(E)$; hence $\Omega_{N_0}(E) \subset \mathbf{T}^2$ may be covered by at most $N_0^C \varepsilon^{-1}$ discs of radius $\varepsilon$. Given $a, x \in \mathbf{T}^2$, estimate next

$$\#\{n = 1, \ldots, L \big| \|T^n x - a\| < \varepsilon\} \tag{15.20}$$

where

$$\|T^n x - a\| = \|x_1 + nx_2 + \frac{n(n-1)}{2}\omega - a_1\| + \|x_2 + n\omega - a_2\|$$

($\| \ \|$ refers to the distance on $\mathbf{T}$ or $\mathbf{T}^2$).

**Lemma 15.21.** *Assume that $\omega$ satisfies (15.0), and let $\varepsilon > L^{-1/10}$. Then*

$$\#\{n = 1, \ldots, L \big| \|T^n x - a\| < \varepsilon\} < C\varepsilon^2 L$$

**Proof.** Majorizing $\chi_{\Delta(a,\varepsilon)}$ by a trigonometric polynomial of degree $< \frac{1}{\varepsilon}$, (15.20) may be bounded by the expression

$$\varepsilon^2 \left( L + \sum_{k \in \mathbf{Z}^2, 0 < |k| < \frac{1}{\varepsilon}} \left| \sum_{n=1}^{L} e^{2\pi i k . T^n x} \right| \right) \tag{15.22}$$

where

$$\left| \sum_{n=1}^{L} e^{2\pi i k . T^n x} \right| = \left| \sum_{n=1}^{L} \exp 2\pi i \left[ (k_1 x_2 + k_2 \omega) n + k_1 \omega \frac{n(n-1)}{2} \right] \right| \tag{15.23}$$

If $k_1 = 0$, we get from (15.0)

$$(15.23) \leq \frac{1}{\|k_2 \omega\|} < C k_2^2 \tag{15.24}$$

For $k_1 \neq 0$, (15.23) is a Gauss-sum that we bound the usual way. Thus

$$(15.23)^2 \quad \leq C \sum_{m \leq L} \frac{1}{\frac{1}{L} + \|(k_1 x_2 + k_2 \omega) + k_1 \omega m\|}$$

$$\overset{\text{by}(15.10)}{\leq} C \sum_{s \mid \frac{1}{L} \leq 2^{-s} \leq 1} 2^s \left( \frac{L}{2^{s/2} |k_1|^{-1}} + 1 \right)$$

$$\leq C |k_1| . L^{3/2}$$

and

$$(15.23) < C |k_1|^{1/2} L^{3/4} \tag{15.25}$$

Therefore, by the choice of $\varepsilon$,

$$(15.22) < C\varepsilon^2 \left( L + \sum_{|k_2| < \frac{1}{\varepsilon}} k_2^2 + L^{3/4} \frac{1}{\varepsilon} \sum_{|k_1| < \frac{1}{\varepsilon}} |k_1|^{1/2} \right)$$

$$< \varepsilon^2 \left( L + \frac{1}{\varepsilon^3} + L^{3/4} \frac{1}{\varepsilon^{5/2}} \right)$$

$$< \varepsilon^2 L$$

This proves the lemma.

Returning to $\Omega_{N_0}(E)$, Lemma 15.21 implies that for arbitrary $x \in \mathbf{T}^2$,

$$\#\{n = 1, \ldots L | \; T^n x \in \Omega_{N_0}(E)\} < N_0^C \varepsilon^{-1} (\varepsilon^2 L) = N_0^C \varepsilon L$$

provided $\varepsilon > L^{-1/10}$. Taking $\varepsilon = L^{-\frac{1}{10}}, L > N_0^{20C}$, we therefore obtain (15.13) with $\rho = \frac{19}{20}$. The condition $L > N^{\frac{1-\rho}{4}}$ becomes $L > N^{\frac{1}{80}}$. We let $N_0 = N^{10^{-10}/C}, N_1 = N^{10^{-4}/C}$ so that $N_0 < N_1^{10^{-3}(1-\rho)^2}$ $\left( \text{see} (14.36) \right)$. From (14.29), $\text{mes}\, \Omega_N(E) < e^{-N^{10^{-2}(1-\rho)^2}} < e^{-N^{10^{-5}}}$ and we may let $\sigma = 10^{-5}$ in (15.11).

This proves Proposition 15.8.

The proof of localization and dynamical localization then follows the same strategy as explained in Chapter 10. The key ingredients are Proposition 15.8 and the following fact:

**Lemma 15.26.** *Let $S \in \mathbf{T}^3$ be a semialgebraic set of degree $B$ s.t.*

$$\text{mes}\, S < e^{-B^\sigma} \quad \text{for } \sigma > 0 \tag{15.27}$$

*Let $M$ be an integer satisfying*

$$\log\log M \ll \log B \ll \log M \qquad (15.28)$$

*Thus, for any fixed $x_0 \in \mathbf{T}^2$,*

$$\mathrm{mes}[\omega \in \mathbf{T}\,|\,(\omega, T_\omega^j x_0) \in S \text{ for some } j \sim M] < M^{-c} \qquad (15.29)$$

*for some $c > 0$.*

*($T_\omega$ denotes the skew shift with frequency $\omega$.)*

**Proof.** For notational reasons, denote $(x_1, x_2) \in \mathbf{T}^2$ by $(x, y)$. The issue is thus the intersection of $S \subset [0, 1]^3$ and sets

$$\{(\omega, x_0 + jy_0 + \frac{j(j-1)}{2}\omega, y_0 + j\omega)|\omega \in [0, 1]\} \qquad (15.30)$$

where $x_0 + jy_0 + \frac{j(j-1)}{2}\omega$ and $y_0 + j\omega$ are considered (mod 1).

We first apply Lemma 9.9 to be set $S$ (with $\eta = e^{-B^\sigma}$) in the product $[0, 1] \times [0, 1]^2$ (thus $\omega \in [0, 1], (x, y) \in [0, 1]^2$). Take

$$\varepsilon = M^{-1+}$$

and consider the decomposition $S = S_1 \cup S_2$ satisfying (9.10) and (9.11). Since $\mathrm{Proj}_\omega S_1$ has measure $< B^C M^{-1+} = M^{-1+}$, restriction of $\omega$ permit us to replace $S$ by $S_2$, satisfying

$$\mathrm{mes}_2(S_2 \cap L) < B^C \varepsilon^{-1}\eta^{1/3} < \eta^{1/4}$$

whenever $L$ is a plane satisfying $|\mathrm{Proj}_L(e_\omega)| < \frac{\varepsilon}{100}$.

Fixing $j$, notice next that (15.30) considered as a subset of $[0, 1]^3$ lies in the union of the parallel planes

$$\mathfrak{S}_m^{(j)} = \mathfrak{S}_m = \left[\omega = \frac{y}{j}\right] - \frac{m + y_0}{j}e_0 \quad (m \in \mathbf{Z}, |m| < M)$$

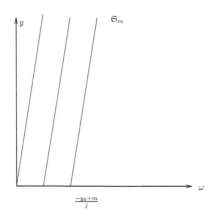

Thus $\mathfrak{S}_m \perp \zeta_j // (1, 0, -\frac{1}{j}), j \sim M$ and $|\zeta_j - e_0| \lesssim M^{-1} \ll \varepsilon$. Hence, for each $m$,

$$\mathrm{mes}(S_2 \cap \mathfrak{S}_m) < \eta^{1/4} \qquad (15.31)$$

Fixing $m$, consider the semialgebraic set $S_2 \cap \mathfrak{S}_m$ and its intersection with the parallel lines

$$\mathcal{L}_{m,m'}^{(j)} = \mathcal{L}_{m,m'}$$
$$= \mathfrak{S}_m \cap \left[ y = \tfrac{2}{j-1} x - \tfrac{2}{j-1} (x_0 + \tfrac{j+1}{2} y_0 - m') \right] \quad (m' \in \mathbf{Z}, |m'| < M)$$

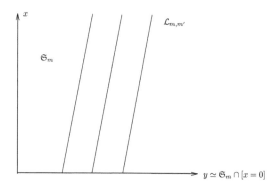

Apply again Lemma 9.9 in the $\mathfrak{S}_m$-plane ($n = 1$) considering the set $S_2 \cap \mathfrak{S}_m (x_0 = y, x_1 = x)$, and let $\varepsilon = M^{-1+}$. The set $S_2 \cap \mathfrak{S}_m$ decomposes as

$$S_2 \cap \mathfrak{S}_m = S_m' \cup S_m'' \quad \text{(depending on } j)$$

where

$\text{Proj}_{\mathfrak{S}_m \cap [x=0]}(S_m')$ is a union of at most $B^C$ intervals of measure at most $B^C M^{-1+}$
(15.32)

and, for all $m'$, by (15.31)

$$\text{mes}\,(S_m'' \cap \mathcal{L}_{mm'}) < B^C M \,\text{mes}\,(S_2 \cap \mathfrak{S}_m)^{1/2} < B^C M \eta^{1/8} \qquad (15.33)$$

Summing (15.33) over $j, m, m'$, the collected contribution in the $\omega =$ parameter is thus less than

$$\frac{1}{M^2} M^3 B^C M \eta^{1/8} < \eta^{1/9}$$

Consider the contribution of the sets $S_m'$. Thus

$$\delta_{j,m} \equiv \text{mes}_\omega \left( \text{Proj}_\omega \text{Proj}_{\mathfrak{S}_m \cap [x=0]}(S_m') \right) < B^C M^{-2+} \qquad (15.34)$$

If the statement (15.29) fails, we have thus

$$\sum_{\substack{j \sim M \\ |m| < M}} \delta_{j,m} > M^{0-} \qquad (15.35)$$

The preceding shows that this is only possible if there is a set $\mathcal{J} \subset \mathbf{Z} \cap [j \sim M], |\mathcal{J}| > M^{1-}$ such that for each $j \in \mathcal{J}$ there are at least $M^{1-}$ values of $m$ satisfying

$$\sum_{|m'| < M} \text{mes}\,\text{Proj}_{\mathfrak{S}_m \cap [x=0]}(S_m' \cap \mathcal{L}_{mm'}^{(j)}) > M^{-1+}$$

Since $\mathrm{Proj}_{\mathfrak{S}_m \cap [x=0]}(S'_m)$ satisfies (15.32), it clearly follows that $S'_m \cap \mathcal{L}_{mm'} \neq \phi$ for at most $M^{0+}$ values of $m'$. Hence

$$\max_{m'} \mathrm{mes}\,(S \cap \mathcal{L}^{(j)}_{m,m'}) > M^{0-} \tag{15.36}$$

For fixed $j$, $\mathcal{L}^{(j)}_{m,m'} \parallel \xi_j, \xi_j \parallel \left(1, \frac{j(j-1)}{2}, j\right)$ and $|\xi_j| = 1$. Since (15.36) holds for at least $M^{1-}$ values of $m$, considerations of semialgebraicity, in particular Bezout's theorem, permit us to find an algebraic curve $\Gamma$ in the $(w, y)$-plane such that the cylindrical surface

$$\mathcal{C}^{(j)} = \Gamma + t\xi_j \tag{15.37}$$

intersects $S_{\eta_1}$ in a set of (2D)-measure $> M^{0-}$. Here $S_{\eta_1}$ denotes the $\eta_1$-neighborhood of $S$, and $\log \eta^{-1} \sim \log \eta_1^{-1}$. Moreover, one may fix $\Gamma$ for $j$ in a set $\mathcal{J}' \subset \mathcal{J}$, with still $|\mathcal{J}'| > M^{1-}$. Denoting again $\mathcal{N}(S, \eta_1)$ the metric entropy numbers, it follows in particular from Theorem 9.4 that

$$\mathcal{N}(\partial S, \eta_1) < B^C \eta_1^{-2} \text{ and } \mathcal{N}(S, \eta_1) < B^C \eta_1^{-2}$$

On the other hand, it follows from the preceding that for $j \in \mathcal{J}'$,

$$\mathcal{N}(S_{\eta_1} \cap \mathcal{C}^{(j)}) > M^{0-} \eta_1^{-2}$$

Therefore,

$$\sum_{j, j' \sim M} \mathcal{N}(\mathcal{C}^{(j)}_{\eta_1} \cap \mathcal{C}^{(j')}_{\eta_1}) > M^{2-} \eta_1^{-2} \tag{15.38}$$

Roughly, (15.38) means that for many pairs $j, j'$ the cylinders $\mathcal{C}^{(j)}, \mathcal{C}^{(j')}$ have $\eta_1$-neighborhoods with large intersections. But the vectors $\{\xi_j\}$ are not coplanar, clearly leading to a contradiction (we use here again (15.28)).

This proves Lemma 15.26.

**Theorem 15.39.** *Consider a lattice operator $H_\omega(x)$ associated to the skew shift $T = T_\omega$ acting on $\mathbf{T}^2$ and of the form (15.2) through (15.6). Fix $x_0 \in \mathbf{T}^2$. Then for almost all $\omega$ satisfying a specified DC and $\delta$ taken sufficiently small in (15.6), $H_\omega(x_0)$ satisfies Anderson localization and dynamical localization.*

**Remarks.**

**1.** The method described in Chapters 14 and 15 is a new perturbative approach to control certain Green's functions developed in [B2] and [BGS2]. It does have a wide range of applications.

**2.** The particular case

$$H_\omega(x) = v(T_\omega x)\delta_{nn'} + \Delta \quad T_\omega x = (x_1 + x_2, x_2 + \omega)$$

may be treated differently, following the method used for the ordinary shift (based on the transfer matrix formalism). But this approach, which we briefly describe below, also has failed so far to produce nonperturbative localization results.

Thus we establish again an LDT for the transfer matrix using the methods from Chapters 4 through 7. This argument is more involved, however, than in the shift case. The method is closely related to arguments in Chapter 7 and is perturbative. The argument may be summarized as follows (see [BGS] for details):

Assume that we dispose of inequalities at scale $N_0$

$$\text{mes}\,[x \in \mathbf{T}^2 \mid \left|\frac{1}{N}\log\|M_N(x,E)\| - L_N(E)\right| > N_0^{-\sigma}] < e^{-N_0^\sigma} \qquad (15.40)$$

for $N = N_0, 2N_0$ and some $\sigma > 0$, where, say,

$$L_{N_0}(E) > 1 \qquad (15.41)$$

and

$$L_{N_0}(E) - L_{2N_0}(E) < \frac{1}{100}L_{N_0}(E) \qquad (15.42)$$

At an initial scale, this may be ensured for $v = \lambda v_0$ ($v_0 = $ a nonconstant real analytic function on $\mathbf{T}^2$) by taking $\lambda$ sufficiently large.

Using (15.40) through (15.42), apply first the avalanche principle with

$$A_j = M_{N_0}(T^{jN_0}x, E) \quad (1 \le j \le n = e^{N_0^{\sigma/2}})$$
$$\mu = e^{N_0} > n$$

Thus

$$|\log\|A_n \cdots A_1\| + \sum_{j=2}^{n-1}\log\|A_j\| - \sum_{j=1}^{n-1}\log\|A_{j+1}A_j\|\,| < C\frac{n}{\mu}$$

and hence, letting $N_1 = nN_0$,

$$\left|\frac{1}{N_1}\log\|M_{N_1}(x,E)\| + \frac{1}{n}\sum_{j=2}^{n-1}\frac{1}{N_0}\log\|M_{N_0}(T^{jN_0}x,E)\| \right. $$
$$\left. - \frac{2}{n}\sum_{j=1}^{n-1}\frac{1}{2N_0}\log\|M_{2N_0}(T^{jN_0}x,E)\|\right| < \frac{C}{\mu} \qquad (15.43)$$

for $x$ outside a set of measure $< 2ne^{-N_0^\sigma}$.

Replacing in (15.43) the element $x$ by each of the elements $\{x, Tx, \ldots, T^{N_0-1}x\}$ and averaging, we get

$$\left|\frac{1}{N_1}\log\|M_{N_1}(x,E)\| + \frac{1}{N_1}\sum_{j=0}^{N_1-1}\frac{1}{N_0}\log\|M_{N_0}(T^jx,E)\| \right.$$
$$\left. - \frac{2}{N_1}\sum_{j=0}^{N_1-1}\frac{1}{2N_0}\log\|M_{2N_0}(T^jx,E)\|\right| \lesssim \frac{N_0}{N_1} \qquad (15.44)$$

for $x$ outside a set of measure $< 2nN_0e^{-N_0^\sigma} < e^{-\frac{1}{2}N_0^\sigma}$.

At this stage, the problem becomes additive, and the dynamics of the skew shift is involved. Assume $v_0$ a trigonometric polynomial for simplicity. Letting

$$u(x) = \frac{1}{N_0}\log\|M_{N_0}(x,E)\|$$
$$= \frac{1}{N_0}\log\left\|\prod_{N_0}^{1}\begin{pmatrix} v(x_1 + jx_2 + \frac{j(j-1)}{2}w, x_2 + jw) - E & -1 \\ 1 & 0 \end{pmatrix}\right\|$$

the function $u = u(x_1, x_2)$ admits a pluri-subharmonic extension $\tilde{u} = \tilde{u}(\tilde{x}_1, \tilde{x}_2)$, $|\operatorname{Im} \tilde{x}_1| \leq 1$, $|\operatorname{Im} \tilde{x}_2| \leq 1$, satisfying

$$|\tilde{u}(\tilde{x}_1, \tilde{x}_2)| < \frac{1}{N_0} \log \prod_{j=1}^{N_0} (C + e^{C(\operatorname{Im} \tilde{x}_1| + j(|\operatorname{Im} \tilde{x}_2|))}) < C(1 + |\operatorname{Im} \tilde{x}_1| + N_0 |\operatorname{Im} \tilde{x}_2|)$$

In particular,

$$\left( \sum_{k_2 \in \mathbf{Z}} |\hat{u}(k_1, k_2)|^2 \right)^{1/2} < \frac{C}{|k_1| + 1} \text{ and } \left( \sum_{k_1 \in \mathbf{Z}} |\hat{u}(k_1, k_2)|^2 \right)^{1/2} < \frac{CN_0}{1 + |k_2|}$$
$$(15.45)$$

Straightforward estimates involving Gauss sums enable us to derive from (15.45) and a DC on $\omega$ that

$$\sup_{x_2 \in \Omega_2} \left\| \hat{u}(0) - \frac{1}{N_1} \sum_{j=0}^{N_1-1} u(T^j x) \right\|_{L^1_{x_1}} < N_1^{-c} \qquad (15.46)$$

where $\Omega_2 \subset \mathbf{T}$, $\operatorname{mes}(\mathbf{T} \backslash \Omega_2) < e^{-N_1^c}$ and $c > 0$ depends on $\omega$.

Since for fixed $x_2$ the function

$$\frac{1}{N_1} \sum_{j=0}^{N_1-1} u(T^j x) = \frac{1}{N_1} \sum_{j=0}^{N_1} u(x_1 + jx_2 + \frac{j(j-1)}{2} \omega, x_2 + j\omega)$$

admits a bounded subharmonic extension in $x_1$, it follows from (15.46) and Corollary 4.10 that for $x_2 \in \Omega_2$,

$$\operatorname{mes} [x_1 \in \mathbf{T}| \left| \frac{1}{N_1} \sum_{j=0}^{N_1-1} u(T^j x) - L_{N_0}(E) \right| > N_1^{-c/2}] < e^{-N_1^{c/5}}$$

Hence

$$\operatorname{mes} [x \in \mathbf{T}^2| \quad \left| \frac{1}{N_1} \sum_{j=0}^{N_1-1} u(T^j x) - L_{N_0}(E) \right| > N_1^{-c/2}] <$$
$$\operatorname{mes}(\mathbf{T} \backslash \Omega_2) + e^{-N_1^{c/5}} < 2e^{-N_1^{c/5}}$$

Returning to (15.44), we proved thus that

$$\left| \frac{1}{N_1} \log \|M_{N_1}(x, E)\| + L_{N_0}(E) - 2L_{2N_0}(E) \right| \lesssim \frac{N_0}{N_1} + N_1^{-c/2} \lesssim N_1^{-c/2} \quad (15.47)$$

except for $x$ in a set of measure $< e^{-\frac{1}{2}N_0^\sigma} + 10e^{-N_1^{c/5}} < 2e^{-\frac{1}{2}N_0^\sigma}$.

The function $u_1 = u_1(x_1, x_2) = \frac{1}{N_1} \log \|M_{N_1}(x, E)\|$ admits a pluri-subharmonic extension bound by $CN_1$, and we apply Lemma 4.12 with $\varepsilon_0 \sim N_1^{-c/2}$, $\varepsilon_1 = 2e^{-\frac{1}{2}N_0^\sigma}$, and $B = N_1 < N_0 e^{N_0^{\sigma/2}}$. Thus

$$\operatorname{mes} [x \in \mathbf{T}^2| \left| \frac{1}{N_1} \log \|M_{N_1}(x, E)\| - L_{N_1}(E) \right| > N_1^{-\frac{c}{9}}] <$$
$$\exp -[N_1^{-\frac{c}{9}} + N_1 e^{-\frac{1}{10}N_0^\sigma}]^{-1} < e^{-N_1^{\frac{c}{10}}} \qquad (15.48)$$

where

$$|L_{N_1}(E) + L_{N_0}(E) - 2L_{2N_0}(E)| < 2N_1^{-c/2} \qquad (15.49)$$

The same holds with $N_1$ replaced by $2N_1$.

Thus (15.48) gives (15.40) at scale $M = N_1, 2N_1$, with $\sigma = \frac{c}{10}$. From (15.49),

$$\begin{aligned} L_{N_1}(E) \quad &> 2L_{2N_0}(E) - L_{N_0}(E) - 2N_1^{-c/2} \\ &> L_{N_0}(E) - \tfrac{1}{100}L_{N_0}(E) - 2N_1^{-c/2} \end{aligned} \qquad (15.50)$$

and

$$|L_{N_1}(E) - L_{2N_1}(E)| < 4N_1^{-c/2} \qquad (15.51)$$

**3.** A few comparisons between the 1D operators

$$\lambda \cos(x + n\omega)\delta_{nn'} + \Delta \qquad (15.52)$$

and

$$\lambda \cos\left(x_1 + nx_2 + \frac{n(n-1)}{2}\omega\right)\delta_{nn'} + \Delta \qquad (15.53)$$

(we assume $\omega$ diophantine and make statements $x$ a.e.).

While for the Almost Mathieu operator (15.52) the spectral type depends on $\lambda$ (with transition at $\lambda = 2$) and the spectrum has a Cantor structure (at least proven in certain cases), the "conjectured" behavior for (15.53) is as follows

(i) If $\lambda \neq 0$, then $L(E) > 0$ for all energies

(ii) For all $\lambda \neq 0$, (15.41) has p.p. spectrum with Anderson localization

(iii) There are no gaps in the spectrum.

Based on the weakly mixing properties of the skew shift, the expected behavior is thus that of the random case.

At this point, the known results are

(iv) (15.53) satisfies Anderson localization for $|\lambda|$ sufficiently large (depending on the DC for $\omega$).

(v) For all $\lambda \neq 0$, there is a set of $\omega$'s of positive measure for which mes $\sum_{pp} > 0$.

Statement (iv) follows from Theorem 15.39 and as mentioned, [BGS] contains a different proof.

Statement (v) at least exhibits some differences with the shift model. See [B1] for the proof, which has similarities with the argument in Chapter 9 for the multi-frequency shift.

# *References*

[B1]  J. Bourgain. Estimates on Green's functions, localization and the quantum kicked rotor model, *Annals of Math.* 156(1) (2002), 249–294.

[B2]  J. Bourgain. On the spectrum of lattice Schrödinger operators with deterministic potential, *J. Analyse Math.* 87 (2002), 37–75 and 88 (2002), 221–254.

[GBS]  J. Bourgain, M. Goldstein, W. Schlag. Anderson localization for Schrödinger operators on $\mathbf{Z}$ with potentials given by the skew shift, *Comm. Math. Phys.* 220(3) (2001), 583–621.

[BGS2]  J. Bourgain, M. Goldstein, W. Schlag. Anderson localization on $\mathbf{Z}^2$ with quasi-periodic potential, *Acta Math.* 188 (2002), 41–86.

# Chapter Sixteen

## Application to the Kicked Rotor Problem

We consider the time-dependent Schrödinger equation on $\mathbf{T} = \mathbf{R}/\mathbf{Z}$

$$i\frac{\partial \Psi(t,x)}{\partial t} = a\frac{\partial^2 \Psi(t,x)}{\partial x^2} + ib\frac{\partial \Psi(t,x)}{\partial x} + V(t,x)\Psi(t,x) \qquad (16.1)$$

with potential

$$V(t,x) = \kappa \cos 2\pi x \left( \sum_{n \in \mathbf{Z}} \delta(t-n) \right) \qquad (16.2)$$

corresponding to a periodic sequence of kicks. The monodromy operator $W$ defined by

$$W\psi(t,x) = \Psi(t+1,x) \qquad (16.3)$$

$\left( = \text{time-1 shift under the flow of } (16.1) \right)$ is a unitary operator on $L^2(\mathbf{T})$. In the case of $(16.1)$, $W$ is given by

$$W = U_{a,b} \cdot W_{1,\kappa} \qquad (16.4)$$

where $U_{a,b}$

$$U_{a,b} = e^{i(a\frac{d^2}{dx^2} + ib\frac{d}{dx})} \qquad (16.5)$$

and

$$W_{1,\kappa} = \text{multiplication operator by } e^{i\kappa \cos 2\pi x} \equiv \rho(x) \qquad (16.6)$$

(see [Sin] and [Bel]).

After passing to Fourier transform, $U_{a,b}$ becomes a diagonal matrix

$$U_{ab} = e^{-i(4\pi^2 an^2 + 2\pi bn)}\delta_{mn} \qquad (16.7)$$

and $W_{1,\kappa}$ becomes a Toeplitz matrix

$$W_{1,\kappa}(m,n) = \hat{\rho}(m-n) \qquad (16.8)$$

where one easily verifies that

$$\hat{\rho}(0) = 1 + 0(\kappa^2) \text{ and } |\hat{\rho}(k)| < \sqrt{\kappa}\, e^{-c(\log \frac{1}{\kappa})|k|} \text{ for } k \in \mathbf{Z}\backslash\{0\} \qquad (16.9)$$

Hence, for $\kappa$ small, $W_{1,\kappa}$ is a perturbation of the identity with exponential off-diagonal decay.

Defining

$$H = \frac{1}{2}(W + W^*) \qquad (16.10)$$

we get a self-adjoint operator

$$H_{mn} = \frac{1}{2} e^{-i(4\pi^2 am^2 + 2\pi bm)} \hat{\rho}(m-n) + \frac{1}{2} e^{i(4\pi^2 an^2 + 2\pi bn)} \overline{\hat{\rho}(n-m)} \quad (16.11)$$

that has the format considered in Chapter 15.

Define

$$V(x,y) = \frac{1}{2}(\hat{\rho}(0) e^{-2\pi ix} + \overline{\hat{\rho}(0)} e^{2\pi ix}) \quad (16.12)$$

$$\phi_k(x,y) = \frac{1}{2} \hat{\rho}(k) e^{-2\pi ix} \quad (16.13)$$

by (16.9) satisfying conditions (15.4), (15.5), and (15.6) with $\gamma = \sqrt{\kappa}$. Define

$$\omega = 4\pi a, x_0 = 0, y_0 = b + 2\pi a \quad (16.14)$$

and let $T$ be the skew shift on $\mathbf{T}^2$ with frequency $\omega$. Then

$$T^m(x_0, y_0) = \left( my_0 + \frac{m(m-1)}{2}\omega, y_0 + m\omega \right)$$

$$= (mb + 2\pi m^2 a, b + 2\pi(2m+1)a)$$

Hence, from (16.11),

$$H_{mm} = V\left(T^m(0, y_0)\right) \quad (16.15)$$

$$H_{mn} = \phi_{m-n}\left(T^m(0, y_0)\right) + \overline{\phi_{n-m}}\left(T^n(0, y_0)\right) \quad (16.16)$$

and (15.2) and (15.3) hold (here $T^m(0, y_0)$ refers to its first coordinate).

Thus Theorem 15.39 applies, and fixing initial $(x_0, y_0)$, localization holds for $\omega$ outside a set of small measure. By (16.14), it follows thus in particular that if the parameters $(a, b)$ are restricted outside a set of small measure ($\to 0$ if $\kappa \to 0$), then $H$ and hence $W$ have pure point spectrum with exponentially decaying eigenfunctions $\{\varphi_\alpha\}$. These eigenfunctions satisfy, moreover, the following property (see (10.37)):

$$\sum_\alpha \left( \sum_n (1+n^2)|\varphi_\alpha(n)|^2 \right)^{1/2} |\langle \psi, \varphi_\alpha \rangle| < \infty \quad (16.17)$$

assuming

$$|\psi_n| < |n|^{-A} \text{ for } |n| \to \infty \quad (16.18)$$

(for $A$ a sufficiently large constant).

Writing

$$W\varphi_\alpha = e^{i\theta_\alpha} \varphi_\alpha$$

we have for $r \in \mathbf{Z}_+$

$$W^r \psi = \sum e^{ir\theta_\alpha} \langle \varphi_\alpha, \psi \rangle \varphi_\alpha$$

and thus

$$\|\Psi(r, x)\|_{H^1(\mathbf{T})} = \left( \sum |n|^2 |(W^r\psi)(n)|^2 \right)^{1/2}$$

$$\leq \sum_\alpha |\langle \varphi_\alpha, \psi \rangle| \left( \sum_n |n|^2 |\varphi_\alpha(n)|^2 \right)^{1/2} < \infty$$

Hence $\psi(t)$ is almost periodic and uniformly bounded in time as an $H^1$-valued map.

The conclusion is the following:

**Theorem.** *Consider the kicked rotor Schrödinger equation (16.1) with small $\kappa$. Then, for $(a, b)$ outside a set of small measure, the following property holds: Let $\Psi(t, x)$ satisfy (16.1) and $\Psi_0 = \Psi(0, x)$ a smooth function on $\mathbf{T}$, more precisely $\Psi_0 \in H^A(\mathbf{T})$, i.e.,*

$$\left( \sum |\hat{\Psi}_0(j)|^2 (1 + j^2)^A \right)^{1/2} < \infty \qquad (16.19)$$

*where $A$ is a sufficiently large constant. Then $\Psi$ is an almost periodic function of time, say, as $H^1(\mathbf{T})$-valued map and*

$$\sup_{t \in \mathbf{R}} \| \Psi(t) \|_{H^1(\mathbf{T})} < \infty \qquad (16.20)$$

Of course, $H^1$ may be replaced by any $H^s$-space (with fixed $s$) provided $A = A(s)$ is (16.19) chosen large enough.

**Remarks.**

**1.** In the previous theorem, one may in fact fix any $b$ and restrict the parameter $a$ outside a set of small measure.

**2.** The classical kicked rotor model has two-dimensional phase space with canonical variables ($\theta$ = rotation angle, $L$ = angular momentum) and time-dependent Hamiltonian

$$H = \frac{1}{2I} L^2 + BL + k \cos \theta \left[ \sum_{n \in \mathbf{Z}} \delta(t - n) \right] \qquad (16.21)$$

($I$ = inertia, $B$ uniform magnetic field).

The Hamilton-Jacobi equations are thus

$$\begin{cases} \frac{d\theta}{dt} = \frac{\partial H}{\partial L} = \frac{L}{I} + B \\ \frac{dL}{dt} = -\frac{\partial H}{\partial \theta} = k \sin \theta \left[ \sum \delta(t - n) \right] \end{cases}$$

Denoting $\theta_n = \theta(n-), A_n = \frac{L(n-)}{I} + B$, we obtain the following equations on $\mathbf{T} \times \mathbf{R}$:

$$\begin{cases} \theta_{n+1} = \theta_n + A_{n+1} \\ A_{n+1} = A_n + \frac{k}{I} \sin \theta_n \end{cases}$$

Corresponding to the Chirikov standard map,

$$\begin{cases} \theta' = \theta + A + \frac{k}{I} \sin \theta \\ A' = A + \frac{k}{I} \sin \theta \end{cases} \qquad (16.22)$$

where $k$ is the coupling constant.

If $k$ is small, most (but not all) of the phase space is filled with (KAM) quasi-periodic motion. It is conjectured that for large $k$ in a set of density $\to 1$ for $k \to \infty$, the system is globally chaotic with no KAM-islands. The quantum version involves $L^2(\mathbf{T}, \frac{d\theta}{2\pi})$ as (infinite-dimensional) phase space and is obtained by replacing $L$ with the operator $i \frac{\partial}{\partial \theta}$. It was introduced by Casati, Chirikov, Izraelev, and Ford

[CCFI] as a quantum analogue of the usual standard map. Thus we consider the time-dependent Schrödinger equation

$$i\frac{\partial \psi}{\partial t} = H(t)\psi$$
$$= -\frac{1}{2I}\frac{\partial^2 \psi}{\partial \theta^2} + iB\frac{\partial \psi}{\partial \theta} + k\cos\theta\left(\sum_{n\in\mathbf{Z}}\delta(t-n)\right)\psi$$

which is equation (16.1). The dynamics in the quantum case are different. Indeed, the theorem above implies (at least for small $k$), typically, the absence of chaotic motion in the quantum model (i.e., the quantum suppression of chaos).

**3.** In the literature, the study of the monodromy operator of the kicked rotor

$$W = e^{i\left(a\frac{d^2}{dx^2} + ib\frac{d}{dx}\right)}\, e^{i\kappa\cos x}$$

is reduced to localization issues for a self-adjoint lattice operator of the form (see [FGP])

$$tg(\gamma n^2 + a)\delta_{nn'} + S_\phi \tag{16.23}$$

where $S_\phi$ is a Toeplitz operator with rapid off-diagonal decay.

This reduction is different from ours and leads to an operator with simpler form. However, one has to deal with the singularity of the $tg$ function. It is likely that the method explained in Chapters 14 and 15 also may be adapted to establish localization results for (16.23).

**4.** If in (16.1) we would consider a randomly kicked model, the behavior would be different, and chaotic diffusion would occur.

As an example, consider the equation

$$i\psi_t = \partial_x^2\psi + V(x,t)\psi$$

with

$$V(x,t) = V_\sigma(x,t) = \kappa\left[\cos 2\pi x\left(\sum_{n\in\mathbf{Z}}\sigma'_n\delta(t-n)\right) + \sin 2\pi x\left(\sum\sigma''_n\delta(t-n)\right)\right] \tag{16.24}$$

and $\sigma = (\sigma', \sigma'')$ as sequences of independently chosen $\pm 1$ signs. We claim that if $\psi_0 = \psi(0) \in H^1(\mathbf{T})$, $\psi_0 \neq 0$, then

$$\varlimsup_{t\to\infty}\frac{\|\psi(t)\|_{H^1}}{t^{1/2}} > 0 \quad \sigma \text{ a.s.} \tag{16.25}$$

Indeed, we have

$$\psi(n+1-) = e^{i\frac{d^2}{dx^2}}\, e^{i(\sigma'_n\cos 2\pi x + \sigma''_n\sin 2\pi x)}\,\psi(n-)$$
$$\|\psi(n+1-)\|_{H^1} = \|e^{i(\sigma'_n\cos 2\pi x + \sigma''_n\sin 2\pi x)}\,\psi(n-)\|_{H^1}$$
$$= \|\partial_x\psi(n-) + 2\pi i(-\sigma'_n\sin 2\pi x + \sigma''_n\cos 2\pi x)\psi(n-)\|_{H^1} \tag{16.26}$$

where $\psi(n-)$ depends only on $\sigma'_j, \sigma''_j$ for $j < n$.

Hence, denoting

$$I_n = \int \|\psi(n-)\|_{H^1}^2\, d\sigma$$

it follows from (16.26) and conservation of the $L^2$-norm that

$$I_{n+1} = \int\int[|\partial_x\psi(n-)|^2 + 4\pi^2(\sin^2 2\pi x + \cos^2 2\pi x)|\psi(n-)|^2]dx d\sigma$$
$$= I_n + 4\pi^2\|\psi(0)\|_2^2$$

This proves (16.25).

**5.** The previous example involves a nonsmooth potential. The following general result holds in the smooth case (see [B2]).

**Theorem.** *Consider the linear Schrödinger equation*

$$i\psi_t + \Delta\psi + V(x,t)\psi = 0$$

*where $x \in \mathbf{T}^d$ (d arbitrary) and $V(x,t)$ is a bounded, real and smooth potential (bounds are uniform in time), periodic in $x$ (no specified behavior in t). If $\psi_0 \in H^s(\mathbf{T}^d)$, then for all $\varepsilon > 0$*

$$\|\psi(t)\|_{H^s} \ll t^\varepsilon \quad \text{for } t \to \infty$$

Thus, if diffusion to higher Fourier modes occurs when $t \to \infty$, it necessarily happens slowly. Such slow diffusion may occur in $H^s$ for any $s > 0$ even if $V$ is smooth and periodic in both $x$ and $t$ (see [B1]). Notice that in the kicked rotor result we did exploit the presence of parameters ($a$ and $b$) to establish absence of diffusion.

# *References*

[Bel]  J. Bellissard. Noncommutative methods in semiclassical analysis, *Springer LNM 1589* (1994), 1–64.

[B1]  J. Bourgain. Growth of Sobolev norms in linear Schrödinger equations with quasi-periodic potential, *CMP* 204 (1999), 207–247.

[B2]  J. Bourgain. On the growth of Sobolev norms in linear Schrödinger equations with smooth time potential, *J. Analyse* 72 (1999), 289–310.

[CCFI]  G. Casati, B. Chirikov, J. Ford, F. Izraelev. *Lecture Notes in Physics*, pp. 334–352, Springer-Verlag, Berlin, 1993.

[CIS]  B. Chirikov, F. Izraelev, D. Shepelyanskii. Dynamical stochasticity in classical and quantum mechanics, *Math. Phys. Reviews* 2 (1981), 209–267.

[FGP]  S. Fishman, D. Grempel, R. Prange. Chaos, quantum recurrences, and Anderson localization, *Phys. Rev. Letters* 49 (1982), 509–512.

[Sin]  Y. Sinai. Mathematical problems in the theory of quantum chaos, *Springer LNM* 1469 (1991), 41–59.

# Chapter Seventeen

## Quasi-Periodic Localization on the $\mathbf{Z}^d$-lattice $(d > 1)$

Consider quasi-periodic $d$-dimensional lattice Schrödinger operators

$$H_\lambda(x) = \lambda v(x_1 + n_1\omega_1, \ldots, x_d + n_d\omega_d)\delta_{nn'} + \Delta \quad (n \in \mathbf{Z}^d) \qquad (17.1)$$

with $\Delta$ the lattice Laplacian on $\mathbf{Z}^d$, i.e.,

$$\Delta(n, n') = 1 \ \text{ if } \sum |n_j - n'_j| = 1$$
$$= 0 \ \text{ otherwise}$$

We assume $v$ a trigonometric polynomial or real analytic function on $\mathbf{T}^d$. More generally, one may consider operators of the form

$$H(x) = \lambda v(x_1 + n_1\omega_1, \ldots, x_d + n_d\omega_d)\delta_{nn'} + S_\phi \qquad (17.2)$$

where $S_\phi(n, n') = \hat{\phi}(n - n')$ is a Toeplitz operator with real analytic symbol. We always assume $\omega \in \mathbf{T}^d$ diophantine.

On the $\mathbf{Z}^d$-lattice, $d > 1$, the transfer matrix approach to localization is not available, and in fact, all our results are perturbative. They will be obtained by adaptation of the method developed in Chapters 14 and 15 to the $d$-dimensional setting.

The main result is the following (see [BGS]):

**Theorem 1.** *Take $d = 2$ and assume $v$ real analytic such that, moreover, none of the partial maps $v(x_1, \cdot), v(\cdot, x_2)$ is a constant function. Then, for all $\varepsilon > 0$, $x \in \mathbf{T}^2$, and $\lambda > \lambda_0(\varepsilon)$, there is a set of frequencies $\mathcal{F}_\varepsilon \subset \mathbf{T}^2$, $\mathrm{mes}(\mathbf{T}^2 \backslash \mathcal{F}_\varepsilon) < \varepsilon$, such that if $\omega \in \mathcal{F}_\varepsilon$, the operator $H_\lambda(x)$ satisfies Anderson localization.*

This result was proven in [BGS] for operators of the form (17.1), and the argument may be extended to case (17.2) as well.

One may consider in particular a potential $v$ obtained as

$$v(x_1, \ldots, x_d) = w(x_1 + x_2 + \cdots + x_d) \qquad (17.3)$$

with $w$ nonconstant real analytic on $\mathbf{T}$. This situation is considerably simpler because the dynamics happen on $\mathbf{T}$, and localization results were obtained in [CD] for this particular class ($d$ is arbitrary). Thus the next result is not new (the proof sketched below differs from [CD], however).

**Theorem 2.** *Consider the Schrödinger operator*

$$H_\lambda = \lambda w(n.\omega)\delta_{nn'} + \Delta \qquad (17.4)$$

*or, more generally,*

$$H_\lambda = \lambda w(n\omega)\delta_{nn'} + S_\phi \tag{17.5}$$

on the $\mathbf{Z}^d$-lattice $(d \geq 1)$. *Here $w$ is a nonconstant real analytic function on* $\mathbf{T}$, *and $\phi$ a real analytic function on $\mathbf{T}^d$. Then, for all $\varepsilon > 0$, there is a set of frequencies $\mathcal{F}_\varepsilon \subset \mathbf{T}^d$ s.t. mes$(\mathbf{T}^d \backslash \mathcal{F}_\varepsilon) < \varepsilon$, and $H_\lambda$ satisfi es Anderson localization for $\omega \in \mathcal{F}_\varepsilon, \lambda > \lambda_0(\varepsilon)$.*

**Remarks.**

**1.** Operators of the form (17.5) come up naturally in several instances. If we consider the 1D multifrequency Schrödinger operator (on $\mathbf{Z}$)

$$H(x) = \lambda\phi(x_1 + n\omega_1, \ldots, x_d + n\omega_d)\delta_{nn'} + \Delta \tag{17.6}$$

the dual operator according to the Aubry duality is indeed

$$\widetilde{H}(\theta) = 2\cos 2\pi(\theta + n\omega)\delta_{nn'} + \lambda S_\phi \tag{17.7}$$

on the $\mathbf{Z}^d$-lattice. In particular, if $\xi \in \ell^2(\mathbf{Z})$ is an eigenstate of $H(x)$,

$$\big(H(x) - E\big)\xi = 0$$

then $(\xi_n)_{n \in \mathbf{Z}^d}$ given by

$$\zeta_n = e^{2\pi i n x} \sum_{j \in \mathbf{Z}} \xi_j e^{2\pi i j(\theta + n.\omega)}$$

defines $(\theta$ a.s$)$ an extended state of $\widetilde{H}(\theta)$ with energy $E$.

Conversely, if $\zeta \in \ell^2(\mathbf{Z}^d)$ is an eigenstate of (17.7), then $(\xi_j)_{j \in \mathbf{z}}$ given by

$$\xi_j = e^{ij\theta} \sum_{n \in \mathbf{Z}^d} \zeta_n\, e^{2\pi i n(x + j\omega)}$$

defines $(x$ a.s.$)$ an extended state of $H(x)$.

Moreover

$$\operatorname{Spec} H = \operatorname{Spec} \widetilde{H}$$

It follows from Theorem 2 that for $\lambda$ small enough, (17.7) typically has only point spectrum, and hence (17.6) has only continuous spectrum (this statement, in fact, uses Green's function estimates on $\widetilde{H}$ that only require $\omega$ to be diophantine). Thus for diophantine $\omega$ in (17.6) and $\lambda$ sufficiently small, (17.6) has only continuous spectrum. Observe that, in view of Theorem 13.3, this statement (and hence also Theorem 2) is perturbative, and the $\lambda$-smallness condition depends on $\omega$. Let $E$ be an eigenvalue of $\widetilde{H}(\theta)$, and let $L(E)$ denote the Lyapounov exponent of $H$. Then, necessarily, $L(E) = 0$. Hence, by continuity of the Lyapounov exponent (see Corollary 7.17), $L(E) = 0$ on $\sum_{pp} \widetilde{H}(\theta) = \operatorname{Spec} \widetilde{H} = \operatorname{Spec} H$. Moreover, adjustment of the proof of Proposition 12.14 to the multidimensional case permits us to show that mes $(\operatorname{Spec} \widetilde{H}) > 0$. Thus mes $[L(E) = 0] > 0$, implying by Kotani's theorem (Proposition 12.5) that $\sum_{ac} H(x) \neq \phi$ for $\omega \in \Omega_\lambda \subset \mathbf{T}^d$ (where mes $(\mathbf{T}^d \backslash \Omega_\lambda) \to 0$ for $\lambda \to 0$) and $x$ a.s.

**2.** As we will explain below, our proof of Theorem 1 in [BGS] does involve some additional difficulties (compared with the $D = 1$ case), and those were only taken care of for $d = 2$ at this point.

The proof of Theorem 1 does involve the same strategy as in the 1D case. The first result needed deals with Green's function estimates for fixed energy and at various scales $N$. Thus, let

$$G_N(E, x) = [H_{[0,N]^2}(x) - E]^{-1}$$

Our aim is then to obtain the usual bounds

$$\|G_N(E, x)\| < e^{N^{1-}} \tag{17.8}$$

$$|G_N(E, x)(n, n')| < e^{-c|n-n'|} \text{ for } |n - n'| > \frac{N}{10} \tag{17.9}$$

for all $x \in \Omega_N(E) \subset \mathbf{T}^2$ with small complement. In the present case, we require estimates of the form

$$\text{mes}\,[x_1' \in \mathbf{T} | (x_1', x_2) \notin \Omega_N(E)] < e^{-N^c}$$

$$\text{mes}\,[x_2' \in \mathbf{T} | (x_1, x_2') \notin \Omega_N(E)] < e^{-N^c} \tag{17.10}$$

for any fixed $x_1, x_2 \in \mathbf{T}$ (which is thus a stronger statement than $\text{mes}\,\Omega_N(E) < e^{-N^c}$).

This fact will be established following a multiscale procedure according to the method explained in Chapters 14 and 15. There are a number of rather obvious modifications when treating the 2D case. For instance, neighborhoods of sites will now be 2D squares; for technical reason, when exploiting the resolvent identity, one needs, however, to enlarge this class a bit to "fundamental regions." These include, besides squares, also differences of squares

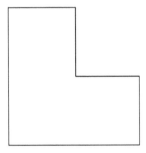

(see [BGS] for details).

A more serious issue and the main difficulty here has to do with the bound

$$\#\{n \in \mathbf{Z}^2 | \,|n_1| + |n_2| < N \text{ and } (x_1 + n_1\omega_1, x_2 + n_2\omega_2) \notin \Omega_{N_0}(E)\} < N^b \tag{17.11}$$

for some $b < 1$ and $\log N_0 \ll \log N$. This is a key element in the method (and the only difficulty for generalizing to arbitrary dimension).

Ensuring this fact will require us to make a new arithmetic specification of $\omega$, which is more involved than the usual diophantine conditions (also generically–in the measure sense–satisfied). This condition is explained in Lemma 17.12 below.

Observe that $\mathcal{A} = \Omega_{N_0}(E)$ may be considered semialgebraic of degree at most $N_0^C$ and satisfying, in addition, by (17.10),

$$\text{mes}\,\mathcal{A}_{x_1} < e^{-N_0^c} \text{ and mes}\,\mathcal{A}_{x_2} < e^{-N_0^c} \text{ for all } x_1, x_2 \in \mathbf{T}$$

The relevant lemmas are the following:

**Lemma 17.12.** *Fix a large positive integer $N$. There is a subset $\Omega_N \subset [0, 1]^2$ with*

$$\text{mes}([0, 1]^2 \backslash \Omega_N) < e^{-\sqrt{\log N}} \qquad (17.13)$$

*and such that any $\omega = (\omega_1, \omega_2) \in \Omega_N$ has the following property. Let $q_1, q_1', q_2, q_2' \in$
$\mathbf{Z}$ be bounded in absolute value by $N$ and suppose that the numbers*

$$\begin{cases} \theta_1 = q_1 \omega_1 \quad (\text{mod } 1) \\ \theta_1' = q_1' \omega_1 \\ \theta_2 = q_2 \omega_2 \\ \theta_2' = q_2' \omega_2 \end{cases} \qquad (17.14)$$

*satisfy*

$$|\theta_i|, |\theta_i'| < N^{-1+\delta} \qquad (i = 1, 2) \qquad (17.15)$$

*and*

$$-N^{-3+\delta} < \begin{vmatrix} \theta_1 & \theta_1' \\ \theta_2 & \theta_2' \end{vmatrix} < N^{-3+\delta} \qquad (17.16)$$

*with $\delta$ small enough. Then*

$$\begin{cases} \gcd(q_1, q_1') > N^{1-\delta'} \\ \gcd(q_2, q_2') > N^{1-\delta'} \end{cases} \qquad (17.17)$$

*where $\delta'(\delta) \to 0$ for $\delta \to 0$.*

Observe that the intuitive meaning of Lemma 17.12 is that $\Omega_N$ does not contain too many $\big($see $(17.17)\big)$ small triangles, $\big($see $(17.15)\big)$ of very small area $\big($see $(17.16)\big)$.
Having introduced $\Omega_N$, we may then state

**Lemma 17.18.** *Let $\mathcal{A} \subset [0, 1]^2$ be semialgebraic of degree $B$. Assume*

$$\text{mes}\mathcal{A}_{x_1} < \eta, \text{mes}\mathcal{A}_{x_2} < \eta \text{ for all } x_1, x_2 \in [0, 1] \qquad (17.19)$$

*where $\mathcal{A}_{x_i}$ denotes the section of $\mathcal{A}$. Let*

$$\log B \ll \log N \ll \log \frac{1}{\eta} \qquad (17.20)$$

*Let $\omega \in \Omega_N \subset [0, 1]^2$ be the set introduced in Lemma 17.12. Then, for some $b < 1$,*

$$\#\{(n_1, n_2) \in \mathbf{Z}^2 \mid |n_i| \leq N \text{ and } (n_1 \omega_1, n_2 \omega_2) \in \mathcal{A}(\text{mod } 1)\} < N^b \qquad (17.21)$$

The idea behind Lemma 17.18 is the following: From (17.19), $\mathcal{A}$ has to be very close to an algebraic curve $\Gamma$, and violation of (17.21) would create too many small and nearly flat triangles–impossible by Lemma 17.12.

Once (17.11) is obtained, the Green's function estimates may be derived by the same arguments used in Chapter 14 in a straightforward way. With the Green's function estimates (17.8) through (17.10) at hand, proving localization results is again achieved by the argument from Chapter 10 using semialgebraic set theory. The 1D treatment given in Chapter 10 may be adjusted rather easily to the 2D setting (the semialgebraic set analysis again uses the results from Chapter 9 but is slightly more complicated here; see [BGS]).

**Proof of Lemma 17.12** (sketch). Since by (17.14) and (17.15)

$$|\theta_i| = \|q_i\omega_i\| = |q_i\omega_i - m_i| < N^{-1+}$$
$$|\theta_i'| = \|q_i'\omega_i\| = |q_i'\omega_i - m_i'| < N^{-1+}$$

an initial diophantine restriction of $\omega$ permits us to assume that

$$|q_i|, |q_i'| > N^{1-}$$

Restricting $\omega_i$ to a size $-\frac{1}{N^2}$ interval, the number of pairs $(q, m)$, $|q| < N$, satisfying

$$|q\omega_i - m_i| < N^{-1+}$$

is at most $N^{0+}$.

Consider next the condition (17.16). Write $\omega_i = \omega_{i,0} + \kappa_i$, $|\kappa_i| < \frac{1}{N^2}$, and fix $\omega_{i,0}$. Thus (17.16)

$$-N^{-3+} < \begin{vmatrix} q_1\omega_1 - m_1 & q_1'\omega_1 - m_1' \\ q_2\omega_2 - m_2 & q_2'\omega_2 - m_2' \end{vmatrix} < N^{-3+}$$

has the form

$$|(q_1q_2' - q_1'q_2)\kappa_1\kappa_2 + \alpha_1\kappa_1 + \alpha_2\kappa_2 + \beta| < N^{-3+} \tag{17.22}$$

Assuming $q_1q_2' - q_1'q_2 \neq 0$, (17.22) restricts $\kappa = (\kappa_1, \kappa_2) \in \left[0, \frac{1}{N^2}\right]^2$ to a set of measure at most

$$\frac{N^{-3+}}{|q_1q_2' - q_1'q_2|} \tag{17.23}$$

Fix $\varepsilon > 0$, and assume first

$$|q_1q_2' - q_1'q_2| > N^{1+\varepsilon} \tag{17.24}$$

Then (17.23) $< N^{-4-\varepsilon+}$. Summing over $N^2 \times N^2$ size $\frac{1}{N^2}$-intervals in $[0, 1]^2$ and $N^{0+}$ pairs $(q_i, m_i), (q_i', m_i')$, the total measure contribution in $\omega$-parameter is thus

$$< N^{-4-\varepsilon+}N^4N^{0+} < N^{-\varepsilon+}$$

Next, consider the case

$$|q_1q_2' - q_1'q_2| < N^{1+\varepsilon} \tag{17.25}$$

and assume that

$$\min_{i=1,2} \gcd(q_i, q_i') < R < N^{1-\delta'} \tag{17.26}$$

Estimate the measure of $\omega = (\omega_1, \omega_2)$ for which there are $q_1, q_1', q_2, q_2'$ satisfying

$$\|q_i\omega_i\| < N^{-1+} \quad\quad \|q_i'\omega_i\| < N^{-1+} \tag{17.27}$$

and (17.25) and (17.26).

Let $r_i = \gcd(q_i, q_i')$, $q_i = r_iQ_i, q_i' = r_iQ_i'$, $(Q_i, Q_i') = 1$. From (17.25)

$$|Q_1Q_2' - Q_1'Q_2| < \frac{N^{1+\varepsilon}}{|r_1r_2|} \tag{17.28}$$

Fix $r_1, r_2$. Write

$$\left| \frac{m_i}{q_i} - \frac{m_i'}{q_i'} \right| < N^{-2+}$$

Hence

$$|m_i Q_i' - m_i' Q_i| < \frac{N^{0+}}{|r_i|} \tag{17.29}$$

Observe that

$$\frac{N^{1-}}{|r_i|} < |Q_i|, |Q_i'| < \frac{N}{|r_i|} \tag{17.30}$$

Thus, if $|r_1| \geq |r_2|$, the number of solutions of (17.28) in the range (17.30), since $(Q_i, Q_i') = 1$, is at most

$$\frac{N^2}{r_1^2} \left( 1 + \frac{N^{1+\varepsilon}}{|r_1 r_2|} \right) \left| \frac{r_1}{r_2} \right| N^{0+} < \frac{N^{2+}}{|r_1 r_2|} + \frac{N^{3+\varepsilon+}}{r_1^2 r_2^2}$$

By (17.30) again, the number of $(m_i, m_i')$ in (17.29) is at most $N^{0+}|r_i|$. This gives the following measure bound in $(\omega_1, \omega_2)$-space

$$N^{-4+} N^{0+} |r_1 r_2| \left( \frac{N^{2+}}{|r_1 r_2|} + \frac{N^{3+\varepsilon+}}{r_1^2 r_2^2} \right) = N^{-2+} + \frac{N^{-1+\varepsilon+}}{|r_1 r_2|} \tag{17.31}$$

for fixed $r_1, r_2$. Summing over $(r_1, r_2)$ subject to (17.26), thus $\min(|r_1|, |r_2|) < R < N^{1-\delta'}$, gives

$$N R N^{-2+} + N^{-1+\varepsilon+} < N^{-\delta'+}$$

This proves Lemma 17.12.

**Proof of Lemma 17.18** (sketch). Assume that (17.21) fails, i.e.,

$$\#K > N^{1-}$$

where

$$K = \{(n_1, n_2) \in \mathbf{Z}^2 | \ |n_i| \leq N \text{ and } (n_1 \omega_1, n_2 \omega_2) \in \mathcal{A}(\bmod 1)\}$$

Since mes $\mathcal{A} < \eta$, dist $(x, \partial \mathcal{A}) < \eta^{1/2}$ for all $x \in \mathcal{A}$. Thus, by the uniformization theorem and (17.20), we may assume that

$$\#K_1 > N^{1-} \tag{17.32}$$

with

$$K_1 = \{(n_1, n_2) \in K | \text{ dist } ((n_1 \omega_1, n_2 \omega_2), \Gamma) < \eta^{1/2}\}$$

where "dist" refers to the distance on $\mathbf{T}^2$, and $\Gamma$ is parametrized by

$$\gamma : [0, 1] \to \Gamma, \text{ with } |\gamma'| < 1, |\gamma''| < 1$$

Fix $\delta > 0$. The curve $\Gamma$ may be covered by $N^{1-\delta}$ discs $D_\alpha$ of radius $\frac{1}{N^{1-\delta}}$ and each such disc contains $(n_1 \omega_1, n_2 \omega_2)$ (mod 1) for at most $N^{\delta+}$ pairs $(n_1, n_2) \in K_1$. Therefore, since also Var $(\gamma') < 1$, we may find some $\alpha$ such that the following holds

$$\#K_\alpha > N^{\delta-} \tag{17.33}$$

where

$$K_\alpha = \{n \in K_1 \mid (n_1\omega_1, n_2\omega_2) \in D_\alpha\}$$

and if $P_0, P_1, P_1' \in D_\alpha \cap \Gamma$ are distinct points, then

$$\left| \det \left[ \frac{P_1 - P_0}{|P_1 - P_0|}, \frac{P_1' - P_0}{|P_1' - P_0|} \right] \right| < N^{-1+\delta} \tag{17.34}$$

Observe also that from assumption (17.19) and semialgebraicity

$$\max_{x_2} \#\{n_1 \in \mathbf{Z} \mid |n_1| < N \text{ and } n_1\omega_1 \in A_{x_2} (\bmod 1)\} < CB < N^{0+}$$

$$\max_{x_1} \#\{n_2 \in \mathbf{Z} \mid |n_2| < N \text{ and } n_2\omega_2 \in A_{x_1} (\bmod 1)\} < N^{0+} \tag{17.35}$$

Covering $\Gamma \cap D_\alpha$ by at most $N^{\delta-\delta^2}$ discs of radius $\frac{1}{N^{1-\delta^2}}$, (17.33) and (17.35) permit us to find $\bar{n} \in K_\alpha, n \in K_\alpha$ s.t. $\bar{n}_1 \neq n_1, \bar{n}_2 \neq n_2$, and

$$\|(n_1 - \bar{n}_1)\omega_1\| + \|(n_2 - \bar{n}_2)\omega_2\| < N^{-1+\delta^2} \tag{17.36}$$

Fix $\bar{n}, n \in K_\alpha$, and further let $n' \in K_\alpha$ be a variable site, $n_i' \neq \bar{n}_i$. Define for $i = 1, 2$

$$q_i = n_i - \bar{n}_i \qquad \theta_i = q_i\omega_i \qquad (\bmod 1)$$

and

$$q_i' = n_i' - \bar{n}_i \qquad \theta_i' = q_i'\omega_i \qquad (\bmod 1)$$

satisfying

$$|\theta_1| + |\theta_2| < N^{-1+\delta^2}$$

$$|\theta_1'| + |\theta_2'| < N^{-1+\delta} \tag{17.37}$$

and from (17.34),

$$\left| \begin{array}{cc} \theta_1 & \theta_1' \\ \theta_2 & \theta_2' \end{array} \right| < N^{-1+\delta}(|\theta_1| + |\theta_2|)(|\theta_1'| + |\theta_2'|) < N^{-3+3\delta}$$

Taking $\delta$ small enough, Lemma 17.12 implies (since $\omega \in \Omega_N$) that

$$\gcd(q_i, q_i') > N^{1-\delta'} \text{ for } i = 1, 2$$

Write thus $q_i = r_i Q_i, q_i' = r_i Q_i'$ with $|r_i| > N^{1-\delta'}$; hence $|Q_i| + |Q_i'| \lesssim N^{\delta'}$, $(Q_i, Q_i') = 1$. Take $k_i, k_i' \in \mathbf{Z} \cap [0, N^{\delta'}]$ s.t. $k_i Q_i + k_i' Q_i' = 1$. It follows that

$$\|r_i\omega_i\| = \|(k_i Q_i + k_i' Q_i')r_i\omega_i\| \leq |k_i| \, \|q_i\omega_i\| + |k_i'| \, \|q_i'\omega_i\| < N^{\delta'} N^{-1+\delta} < N^{-\frac{1}{2}}.$$

Therefore,

$$|\theta_i| = \|q_i\omega_i\| = \|Q_i r_i\omega_i\| = |Q_i| \, \|r_i\omega_i\|$$

$$|\theta_i'| = |Q_i'| \, \|r_i\omega_i\|$$

so that

$$\frac{|\theta_i|}{|\theta_i'|} = \frac{|q_i|}{|q_i'|}$$

$$|\theta_i'| = \frac{|q_i'|}{|q_i|}|\theta_i| \le N\frac{|\theta_i|}{|q_i|} \qquad (17.38)$$

Recalling (17.37), $|\theta_i| < N^{-1+\delta^2}$ and $|q_i| > N^{1-\delta^2+}$. Therefore, by (17.38),

$$\|q_i'\omega_i\| = |\theta_i'| < N N^{-2(1-\delta^2)+} < N^{-1+2\delta^2+} \qquad (17.39)$$

But the number of $q_i' \in \mathbf{Z} \cap [0, N]$ satisfying (17.39) is at most $N^{2\delta^2}+r$; hence $K_\alpha - \bar{n}$ contains at most $N^{4\delta^2+}$ elements $n' - \bar{n} = (q_1', q_2')$. This contradicts (17.33) and proves Lemma 17.18.

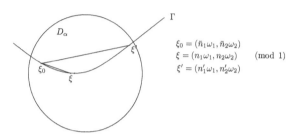

$\xi_0 = (\bar{n}_1\omega_1, \bar{n}_2\omega_2)$
$\xi = (n_1\omega_1, n_2\omega_2) \qquad (\mathrm{mod}\ 1)$
$\xi' = (n_1'\omega_1, n_2'\omega_2)$

Next, a few comments on the proof of Theorem 2, which is easier (in particular there is no need for Lemma 17.12). Define for $\theta \in \mathbf{T}$

$$H(\theta) = \lambda w(n \cdot \omega + \theta) + \Delta$$

and

$$G_N(E, \theta) = [H_{[0,N]^d}(\theta) - E + i0]^{-1}$$

Then the bounds (17.8) and (17.9) are obtained for $\theta \in \Omega_N(E) \subset \mathbf{T}$ with mes $(\mathbf{T}\backslash\Omega_N(E)) < e^{-N^c}$.

For $\log\log N \ll \log N_0 \ll \log N$, the bound (17.11) becomes

$$\#\{n \in [0, N]^d \cap \mathbf{Z}^d \,|\, \theta + n.\omega \notin \Omega_{N_0}(E)\} < N_0^C < N^{0+} \qquad (17.40)$$

Indeed, $\mathbf{T}\backslash\Omega_{N_0}(E)$ is a union of at most $N_0^C$ intervals of size $< e^{-N_0^c}$ and (17.40) only requires a diophantine condition on $\omega$.

## References

[BGS] J. Bourgain, M. Goldstein, W. Schlag. Anderson localization for Schrödinger operators on $\mathbf{Z}^2$ with quasi-periodic potential, *Acta Math.* 188 (2002), 41–86.

[C-Di] V. Chulaevsky, E. Dinaburg. Methods of KAM theory for long-range quasi-periodic operators on $\mathbf{Z}^n$. Pure point spectrum, *Comm. Math. Physics* 153(3) (1993), 559–577.

# Chapter Eighteen

## An Approach to Melnikov's Theorem on Persistency of Nonresonant Lower Dimension Tori

The problem is that of persistency of $b$-dimensional tori in $\mathbf{R}^{2b} \times \mathbf{R}^{2r}$-phase space for a real analytic Hamiltonian $H$ of the form

$$
\begin{aligned}
H = H(I, \theta, y) \; &= H(I_1, \ldots, I_b, \theta_1, \ldots, \theta_b, y_1, \ldots, y_r) \\
&= \langle \lambda_0, I \rangle + \sum_{s=1}^{r} \mu_s |y_s|^2 + |I|^2 + \varepsilon H_1(I, \theta, y)
\end{aligned}
\tag{18.1}
$$

(the last term is perturbative), as considered in [E], [Kuk], [Pos], and [B1]. Here, $I = (I_1, \ldots, I_b), \theta = (\theta_1, \ldots, \theta_b)$ are action-angle variables for the "tangential" part of the phase space, and $y = (y_1, \ldots, y_r)$ are the "normal" coordinates. The vector $\lambda_0$ is diophantine and satisfying certain nonresonance conditions; in [E] and [Kuk], the conditions imposed are

$$
\langle \lambda_0, k \rangle - \mu_s \neq 0 \qquad (k \in \mathbf{Z}^b, s = 1, \ldots, r)
\tag{18.2}
$$

and also

$$
\langle \lambda_0, k \rangle + \mu_s - \mu_{s'} \neq 0 \qquad (s \neq s')
\tag{18.3}
$$

In [B1], the persistency result (as stated in [E]) is obtained assuming only (18.2), which is, perturbatively speaking, the natural condition (and does not exclude multiplicities of the normal frequencies—often present in the PDE context).

We will treat here only the problem of a Hamiltonian perturbation of a linear equation, to which (18.1) may be reduced. Consider thus the equation

$$
\frac{1}{i} \dot{q}_j = \frac{\partial H}{\partial \bar{q}_j} \qquad (1 \leq j \leq B)
\tag{18.4}
$$

with Hamiltonian

$$
H(q, \bar{q}) = \sum_{j=1}^{B} \mu_j |q_j|^2 + \varepsilon H_1(q, \bar{q})
\tag{18.5}
$$

with canonical coordinates $q_1, \ldots, q_B \in \mathbf{C}$. $\{\mu_j\}$ are real, and we assume for simplicity that $H_1(q, \bar{q})$ is given by a polynomial in $q_j, \bar{q}_j$ with real coefficients.

Let $1 \leq b < B$, and consider $\lambda = (\mu_1, \ldots, \mu_b)$ as a $b$-dim parameter (in the context (18.1), these parameters are extracted from the $|I|^2$-term by amplitude-frequency modulation; see [B3]). For simplicity, we assume that the remaining frequencies $\mu_j (b < j \leq B)$ are constant.

Letting $\varepsilon = 0$, (18.4) has the (unperturbed) quasi-periodic solution

$$q_j(t) \begin{aligned} &= a_j e^{i\lambda_j t} & (1 \leq j \leq b) \\ &= 0 & (b < j \leq B) \end{aligned} \qquad (18.6)$$

There is the following persistency result:

**Theorem 18.7.** Let $\lambda$ be restricted to an interval $\Omega \in \mathbf{R}^b$, and assume that $\varepsilon$ in (18.5) is sufficiently small. Fix $a_1, \ldots, a_b \in \mathbf{R}_+$. There is a Cantor subset $\Omega_\varepsilon \subset \Omega$, mes $(\Omega \backslash \Omega_\varepsilon) \overset{\varepsilon \to 0}{\to} 0$, and a smooth map $\lambda \to \lambda'$ on $\Omega$, s.t. for $\lambda \in \Omega_\varepsilon$, and there is a quasi-periodic solution of (18.4)

$$q_j(t) = \sum_{k \in \mathbf{Z}^b} \hat{q}_j(k) e^{ik.\lambda' t} \qquad (1 \leq j \leq B) \qquad (18.8)$$

satisfying

$$\hat{q}_j(e_j) = a_j \qquad (1 \leq j \leq b) \qquad (18.9)$$

$$|\hat{q}_j(k)| < e^{-c|k|} \text{ for some } c > 0 \qquad (18.10)$$

and

$$\sum_{(j,k) \notin \mathcal{R}} |\hat{q}_j(k)| < \sqrt{\varepsilon} \qquad (18.11)$$

where

$$\mathcal{R} = \{(j, e_j) | j = 1, \ldots, b\} \qquad (18.12)$$

This theorem was proven in [B2]. Our aim is to give a different argument based on the methods developed earlier on in this work, such as Chapter 14 and the semialgebraic set techniques. This will enable us to avoid use of the preparation theorem (used in [B2] and [B4]). These preparation theorems may be particularly delicate in PDE applications (involving an infinite-dimensional phase space) due to large numbers of multiplicities in the normal frequencies (see, for instance, [B4] on 2D NLS). In the latter situation, the perturbed polynomials are of unbounded degree and require a precise analysis of the size of allowable perturbations. The problem becomes increasingly difficult for large dimension and was not treated for 3D NLS so far. The present approach avoids the use of preparation theorems altogether and therefore offers much more flexibility.

The method of proof of the theorem is the same as in [B2] and is based on a Lyapounov-Schmidt decomposition in $P$- and Q-equations (used earlier on in [C-W] in this context). The general scheme will be recalled briefly below, and the reader is referred to [B3], for instance, for a detailed exposition. It is well known that the core of the matter in this approach is control of the inverses of certain linear operators. These arise from consecutive linearizations when applying the Newton iteration scheme. The matrices are essentially diagonal with a perturbation given by a Toeplitz-type operator with rapid off-diagonal decay. The diagonal part is quasi-periodic in $k \in \mathbf{Z}^b$ and hence fits the general type discussed earlier.

## I. The Lyapounov-Schmidt Decomposition

Using the Fourier transform, the equations

$$\frac{1}{i}\dot{q}_j = \mu_j q_j - \varepsilon \frac{\partial H_1}{\partial \bar{q}_j} \quad (1 \le j \le B) \tag{18.13}$$

are reformulated on the $\mathbf{Z}^b$-lattice. We denote $(\mu_1, \ldots, \mu_b)$ alternatively by $\lambda = (\lambda_1, \ldots, \lambda_b)$ (= the unperturbed tangential frequencies). Thus $\lambda$ is a parameter; the $\mu_j (b < j \le B)$ are fixed.

Thus (18.3) becomes

$$(k.\lambda' - \mu_j)\hat{q}_j(k) + \varepsilon \widehat{\frac{\partial H_1}{\partial \bar{q}_j}}(k) = 0 \quad \begin{cases} 1 \le j \le B \\ k \in \mathbf{Z}^b \end{cases} \tag{18.14}$$

Specifying as in (18.9),

$$\hat{q}_j(e_j) = a_j \in \mathbf{R}_+ \quad (1 \le j \le b)$$

the $Q$-equations relate $\lambda, \lambda'$

$$\lambda'_j - \lambda_j + \frac{\varepsilon}{a_j} \widehat{\frac{\partial H_1}{\partial \bar{q}_j}}(e_j) = 0 \quad (1 \le j \le b) \tag{18.15}$$

Define

$$u_j(k) = \hat{q}_j(k) \qquad v_j(k) = \widehat{\bar{q}}_j(k) = \overline{u_j(-k)}$$

(these will in fact be real in the present setup and hence the reality condition in (18.15) automatically fulfilled).

Rewrite the system (18.14) as a system in $u$ and $v$

$$\begin{cases} (k.\lambda - \mu_j)u_j(k) + \varepsilon \widehat{\frac{\partial H_1}{\partial \bar{q}_j}}(k) = 0 & (18.16') \\[2mm] (-k\lambda - \mu_j)v_j(k) + \varepsilon \widehat{\frac{\partial H_1}{\partial q_j}}(k) = 0 & (18.16'') \end{cases}$$

where $\widehat{\frac{\partial H_1}{\partial \bar{q}_j}}(k)$ and $\widehat{\frac{\partial H_1}{\partial q_j}}(k)$ are thus multilinear expressions in $u_{j'}(k'), v_{j'}(k')$.

The $P$-equations are obtained by restriction of (18.16') and (18.16'') outside the "resonant" set of sites $\mathcal{R}$, i.e.,

$$(j, k) \notin \mathcal{R} = \{(j, e_j) | j = 1, \ldots, b\} \text{ in } (18.16')$$
$$\notin \mathcal{R}^* = \{(j, -e_j) | j = 1, \ldots, b\} \text{ in } (18.16'')$$

The $P$-equations are solved following a Newton-iteration scheme, leading to a sequence of approximative solutions $\{q_{(r)}\}$, $q_{(r)} = q_{(r)}(\lambda, \lambda')$ depending (smoothly) on $\lambda, \lambda'$. To be an "approximative" solution of the $P$-equations, one needs, however, to make consecutive restrictions of $(\lambda, \lambda')$ (to be explained later). However, $q_{(r)}$ will be defined smoothly on the entire parameter set.

Substitution of the solution of the $P$-equation in $Q$-equations (18.15) will allow us to determine $\lambda' = \lambda'(\lambda)$ by the implicit function theorem. Identifying $q_{(r)}$ and $\hat{q}_r$, supp $q_{(r)}$ will refer in the sequel to the support in Fourier space. For simplicity, denote also the left side of (18.16') and (18.16''), $(j, k) \notin \mathcal{R}$, by $F(q)$.

To perform consecutive corrections $q_{r+1} = q_r + \Delta_{r+1}q$ by Newton's algorithm, one needs to control the linearized operator

$$T_q = D + \varepsilon S_q$$

where $D$ is diagonal

$$\begin{pmatrix} k.\lambda' - \mu_j & 0 \\ 0 & -k\lambda' - \mu_j \end{pmatrix}$$

and

$$S_q = \begin{pmatrix} \widehat{\dfrac{\partial^2 H_1}{\partial \bar{q}_j \partial q_{j'}}}(k - k') & \widehat{\dfrac{\partial^2 H_1}{\partial \bar{q}_j \partial \bar{q}_{j'}}}(k - k') \\[2mm] \widehat{\dfrac{\partial^2 H_1}{\partial q_j \partial q_{j'}}}(k - k') & \widehat{\dfrac{\partial^2 H_1}{\partial q_j \partial \bar{q}_{j'}}}(k - k') \end{pmatrix}$$

Thus the index set is $(\pm, j, k)$ $1 \le j \le B$, $k \in \mathbf{Z}^b$.

$S_q$ is self-adjoint and stationary in $k$-index. It is given by a block matrix formed by Toeplitz operators with rapidly decaying off-diagonal (as consequence of the construction of the $q_{(r)}$).

Following Newton's method, let $q = q_r$,d and define the next increment

$$\Delta_{r+1}q = -T_N^{-1}\big(F(q_r)\big) \tag{18.17}$$

with $N$ referring to the restriction $|k| < N$, with $N$ such that supp $F(q_r) \subset B(0, \frac{N}{10})$. Thus

$$F(q_{r+1}) = (T_N - T)T_N^{-1}\big(F(q_r)\big) + 0\big(\|\Delta_{r+1}q\|^2\big) \tag{18.18}$$

## II. Inductive Assumptions

(i)  supp $q_r \subset B(0, A^r)$

   ($A$ a large constant depending on $H_1$)

(ii)  $\|\Delta_r q\| < \delta_r$

   $\|\partial \Delta_r q\| < \bar{\delta}_r$     ($\partial$ referring to derivation in $\lambda$ or $\lambda'$)

   ($\|\Delta_r q\|$ refers to $\sup_{\lambda, \lambda'} \|\widehat{\Delta_r q}\|_{\ell^2(\mathbf{Z}^b)}$)

**Remark.** The size of $\delta_r, \bar{\delta}_r$ will satisfy $\log \log \frac{1}{\delta_r + \bar{\delta}_r} \sim r$. Only first-order derivatives in $\lambda, \lambda'$ will be needed, but one could include higher-order derivatives as well.

(iii)  $|\hat{q}_r(k)| < e^{-c|k|}$ for some constant $c > 0$

This constant will decrease slightly along the iteration but remain bounded away from 0. With $q_r$ defined as a $C^1$-function on the entire $(\lambda, \lambda')$-parameter space, application of the implicit function theorem to the $Q$-equations

$$\lambda'_j = \lambda_j + \frac{\varepsilon}{a_j} \widehat{\frac{\partial H_1}{\partial \bar{q}_j}}(e_j) \text{ with } q = q_r$$

yields

$$\lambda' = \lambda + \varepsilon \varphi_r(\lambda) \text{ with } \|\partial \varphi_r\| < C$$

which graph we denote by $\Gamma_r$. Clearly, by (ii),

$$|\varphi_r - \varphi_{r-1}| \lesssim \varepsilon \|q_{(r)} - q_{(r-1)}\| < \varepsilon \delta_r$$

so that $\Gamma_r$ is an $\varepsilon \delta_r$-approximation of $\Gamma_{r-1}$.

(iv) There is a collection $\Lambda_r$ of intervals $I$ in $\mathbf{R}^{2b}$ of size $A^{-r^C}$ s.t.

(a) On $I \in \Lambda_r$, $q_r(\lambda, \lambda')$ is given by a rational function in $(\lambda, \lambda')$ of degree at most $A^{r^3}$.

(b) For $(\lambda, \lambda') \in \bigcup_{I \in \Lambda_r} I$,

$$\|F(q_r)\| < \kappa_r$$
$$\|\partial F(q_r)\| < \bar{\kappa}_r \qquad (\partial \text{ refers to } \lambda\text{- or } \lambda'\text{-derivatives})$$

**Remark.** Again, $\log \log \frac{1}{\kappa_r + \bar{\kappa}_r} \sim r$.

(c) For $(\lambda, \lambda') \in \bigcup_{I \in \Lambda_r} I$, $T = T_{q^{(r-1)}}$ satisfies

$$\|T_N^{-1}\| < A^{r^C}$$
$$|T_N^{-1}(k, k')| < e^{-c|k-k'|} \text{ for } |k - k'| > r^C$$

where $T_N$ refers to the restriction $|k| < N = A^r$.

**Remark.** Since $\varepsilon$ is small, the control of $T_N^{-1}$ at initial scales reduces to the diagonal $D$ and hence minoration of the expressions $|k.\lambda \pm \mu_j|, |k| \leq N$.

(d) Each $I \in \Lambda_r$ is contained in an interval $I' \in \Lambda_{r-1}$ and

$$\text{mes}_b \left( \Gamma_r \cap \left( \bigcup_{I' \in \Lambda_{r-1}} I' \setminus \bigcup_{I \in \Lambda_r} I \right) \right) < A^{-r/5} \qquad (18.19)$$

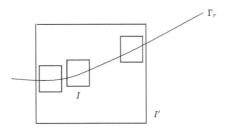

### III. Invertibility of $T_{q^{(r)}}$

Let $N = A^{r+1}$, which may be assumed to be large based on the by previous remark. To construct $\Delta_{r+1} q$, we need to control $(R_N T_{q^{(r)}} R_N)^{-1}$ with a further restriction of the $(\lambda, \lambda')$-parameter set.

Fix $(\lambda, \lambda') \in \bigcup_{I \in \Lambda_r} I$. Since

$$\|T_{q^{(r)}} - T_{q^{(r-1)}}\| \lesssim \|\Delta_r q\| < \delta_r, \quad \log \log \frac{1}{\delta_r} \sim r$$

it follows from the assumption on $R_{A^r} T_{q^{(r-1)}} R_{A^r}$ that $R_{A^r} T_{q^{(r)}} R_{A^r}$ also satisfies the estimate in (c).

Fix $M_0$, satisfying

$$M_0 \sim (\log N)^{C/2} \qquad (18.20)$$

(for some $C$ to be specified).

Assume that the following estimates are obtained for any interval $J \in \mathbf{Z}^b$ of size $M_0$ and centered at some $k \in \mathbf{Z}^b$, $\frac{1}{2} A^r < |k| < A^{r+1}$

$$\|(R_J T_{q^{(r)}} R_J)^{-1}\| < e^{M_0^{1-}} \tag{18.21}$$

$$|(R_J T_{q^{(r)}} R_J)^{-1}(k, k')| < e^{-c|k-k'|} \text{ for } k, k' \in J, |k - k'| > \frac{M_0}{10} \tag{18.21'}$$

It will then follow from the resolvent identity that

$$\|(R_N T_{q^{(r)}} R_N)^{-1}\| < A^{r^C} + e^{M_0^{1-}} \tag{18.22}$$

and

$$|(R_N T_{q^{(r)}} R_N)^{-1}(k, k')| < e^{-c|k-k'|} \text{ for } |k - k'| > r^C \tag{18.22'}$$

To obtain (18.21) and (18.21'), we may perturb $q^{(r)}$ by $0(e^{-M_0})$. Thus we may replace $q^{(r)}$ by $q^{(r_0)}$ with $r_0 < r$ satisfying

$$\delta_{r_0} < e^{-M_0} \quad \text{(hence } r_0 \sim \log M_0) \tag{18.23}$$

Consider the following matrix:

$$T^\sigma = D^\sigma + S_{q^{(r_0)}(\lambda, \lambda')} \tag{18.24}$$

where we introduced an additional (1-dim) parameter $\sigma$ and

$$D^\sigma_{\pm, j, k} = \pm(k.\lambda' + \sigma) - \mu_j \tag{18.25}$$

Hence (18.21) and (18.21') will follow from

$$\|(T^\sigma_{M_0})^{-1}\| < e^{M_0^{-1}} \tag{18.26}$$

and

$$|(T^\sigma_{M_0})^{-1}(k, k')| < e^{-c|k-k'|} \text{ for } |k - k'| > \frac{M_0}{10} \tag{18.26'}$$

valid for all $\sigma \in \{k.\lambda' | \frac{1}{2} A^r < |k| < A^{r+1}\}$.

**Lemma 18.27.** *Assume $\lambda'$ diophantine. Then (18.26) and (18.26') hold for all $\sigma$ outside a set of measure at most $e^{-M_0^{1/2}}$ (and depending on $\lambda, \lambda'$).*

**Remark.** It is straightforward to ensure the required diophantine properties on $\lambda'$ by introducing the interval systems $\Lambda_r$ appropriately.

**Proof of Lemma 18.27.** Follow the multiscale argument from Chapter 14, complexifying in the parameter $\sigma$. The details are straightforward. (Observe that these operators are closely related to those considered in Chapter 17, Theorem 2).

Now fix $I \in \Lambda_{r_0}$, and consider the set

$$\mathfrak{S} = \{(\lambda, \lambda', \sigma) \in I \times \mathbf{R} | (18.26) + (18.26') \text{ fail for } T^\sigma_{M_0}\} \tag{18.28}$$

This condition may be expressed by a polynomial in the matrix elements of $T^\sigma$ of degree $< M_0^C$. These matrix elements are at most linear in $\sigma$ and rational functions in $\lambda, \lambda'$ of degree at most $A^{r_0^3}$ (by assumption (a) on $q^{(r_0)}$). Therefore, $\mathfrak{S}$ is semialgebraic of degree at most $M_0^C A^{r_0^3}$.

Since $\Gamma_{r_0} \cap I$ is defined by an equation in $\lambda, \lambda'$ of degree $< CA^{r_0^3}$, $\mathfrak{S} \cap ((\Gamma_{r_0} \cap I) \times \mathbf{R})$ also is semialgebraic of degree $< M_0^C A^{r_0^3}$. By Lemma 18.27, each $(\lambda, \lambda')$-section is of 1-dim measure $< e^{-M_0^{1/2}}$ so that

$$\operatorname{mes}_{1+b}\big(\mathfrak{S} \cap ((\Gamma_{r_0} \cap I) \times \mathbf{R})\big) < e^{-M_0^{1/2}} \qquad (18.29)$$

Our aim is to estimate for $|k| > \frac{1}{2} A^r$

$$\operatorname{mes}_b\{\lambda | (\lambda, \lambda + \varepsilon\varphi_{r_0}(\lambda), k.(\lambda + \varepsilon\varphi_{r_0}(\lambda))) \in \mathfrak{S} \cap (I \times \mathbf{R})\}$$

Apply first decomposition Lemma 9.9 to the set $\mathfrak{S} \cap ((\Gamma_{r_0} \cap I) \times \mathbf{R})$ as subset of $(\Gamma_{r_0} \cap I) \times \mathbf{R}$ (which may be identified with an interval in $\mathbf{R}^{b+1}$). This gives a subset $\Gamma' \subset \Gamma_{r_0} \cap I$ of measure

$$\operatorname{mes}_b \Gamma' < (M_0 A^{r_0^3})^C A^{-r} < A^{-r/2} \big(\text{ see } (18.20) \text{ and } (18.23)\big) \qquad (18.30)$$

such that for all $k, |k| > \frac{1}{2} A^r$

$$\operatorname{mes}_b\{\lambda | (\lambda, \lambda + \varepsilon\varphi_{r_0}(\lambda), k.(\lambda + \varepsilon\varphi_{r_0}(\lambda))) \in \mathfrak{S} \cap ((\Gamma_{r_0} \setminus \Gamma') \cap I) \times \mathbf{R})\} \quad (18.31)$$
$$< e^{-M_0^{1/3}}$$

Therefore, there is a set $\Gamma'' \subset \Gamma_{r_0} \cap I$

$$\operatorname{mes}_b \Gamma'' < A^{-r/2} + A^{b(r+1)} e^{-M_0^{1/3}} < A^{-r/3}$$

(by choice (18.20) of $M_0$) such that (18.26) and (18.26′) hold for all $(\lambda, \lambda') \in (\Gamma_{r_0} \setminus \Gamma'') \cap I$ and $\sigma = k.\lambda', \frac{1}{2} A^r < |k| < A^{r+1}$. From the previous discussion, this implies (18.21) and (18.21′) on $(\Gamma_{r_0} \setminus \Gamma'') \cap I$.

Letting $I$ range over $\Lambda_{r_0}$, the total measure removed from $\Gamma_{r_0}$ is at most $A^{r_0^C} . A^{-r/3} < A^{-r/4}$. Since (18.21) and (18.21′) allow $0(e^{-M_0})$ perturbation of $(\lambda, \lambda')$ and $\Gamma_{r_0}, \Gamma_r$ are at distance $< \delta_{r_0} < e^{-M_0}$, by (18.23), we obtain a subset $\tilde{\Gamma}_r \subset \Gamma_r, \operatorname{mes}_b \tilde{\Gamma}_r < A^{-r/4}$ such that (18.21) and (18.21′) hold on

$$\bigcup_{I \in \Lambda_{r_0}} \big(I \cap (\Gamma_r \setminus \tilde{\Gamma}_r)\big) \qquad (18.32)$$

and hence on

$$\bigcup_{I \in \Lambda_r} \big(I \cap (\Gamma_r \setminus \tilde{\Gamma}_r)\big) \qquad (18.33)$$

Since on (18.33), as well as (18.22) and 18.22′), holds with $N$ replaced by $A^r$, (18.22) and (18.22′) hold on (18.33). Clearly, since the bound in (18.22) is at most $A^{r^C} + e^{(\log A^{r+1})^C/2} < A^{(r+1)^C}$, (18.22) and (18.22′) remain valid on an $A^{-(r+1)^C}$-neighborhood of (18.33). This gives a collection $\Lambda_{r+1}$ of intervals in $\mathbf{R}^{2b}$ of size $A^{-(r+1)^C}$, s.t. for $(\lambda, \lambda') \in I \in \Lambda_{r+1}$

$$\|(R_{A^{r+1}} T_{q^{(r)}} R_{A^{r+1}})^{-1}\| < A^{(r+1)^C} \qquad (18.34)$$

and (18.22′), and

$$\operatorname{mes}_b\Big(\bigcup_{I \in \Lambda_r}(I \cap \Gamma_r) \setminus \bigcup_{I' \in \Lambda_{r+1}}(I' \cap \Gamma_r)\Big) \le \operatorname{mes}_b \tilde{\Gamma}_r < A^{-r/4} \qquad (18.35)$$

which will imply (18.19) at stage $r + 1$.

**IV. Construction of $q^{(r+1)}$**

Denoting $T = T_{q^{(r)}}$, $N = A^{r+1}$, then define for $(\lambda, \lambda') \in \bigcup_{I \in \Lambda_{r+1}} I$

$$\Delta_{r+1} q = -T_N^{-1}(F(q_r))$$

Thus $\Delta_{r+1} q$ is a rational function in $(\lambda, \lambda')$ of degree at most

$$N^C . A^{r^3} + C A^{r^3} < A^{r^3 + Cr} < A^{(r+1)^3}$$

We have from (18.34) and assumption (b)

$$\|\Delta_{r+1} q\| < A^{(r+1)^C} . \kappa_r = \delta_{r+1} \tag{18.36}$$

and

$$
\begin{aligned}
\|\partial(\Delta_{r+1} q)\| &\leq \|\partial T_N^{-1}\| . \|F(q_r)\| + \|T_N^{-1}\| \, \|\partial F(q_r)\| \\
&< \|T_N^{-1}\|^2 \, \|q_r\|_{C^1} \, \kappa_r + \|T_N^{-1}\| . \bar{\kappa}_r \\
&< A^{2(r+1)^C} \bar{\kappa}_r
\end{aligned} \tag{18.37}
$$

Also,

$$|\widehat{\Delta_{r+1} q}(k)| \leq \sum_{|k'| \leq N} |T_N^{-1}(k, k')| \, |\widehat{F(q_r)}(k')|$$

where by (iii) and (c),

$$|\widehat{F(q_r)}(k')| \leq \sum_{k_1 + \cdots + k_A = k'} |\hat{q}_r(k_1)| \cdots |\hat{q}_r(k_A)| \leq (CA)^{CA} |k'|^{bA} e^{-c|k'|}$$

and

$$
\begin{aligned}
&|\widehat{\Delta_{r+1} q}(k)| \\
&\leq (CA)^{CA} \left\{ \sum_{|k-k'| < r^C} A^{r^C} |k'|^{bA} e^{-c|k'|} + \sum_{|k-k'| \geq r^C} |k'|^{bA} e^{-c(|k'| + |k - k'|)} \right\} \\
&< C' A^{2r^C} |k|^{2bA} e^{-c|k|}
\end{aligned}
$$

On the other hand,

$$\|\Delta_{r+1} q\| < \delta_r, \log \log \frac{1}{\delta_r} \sim r$$

so that the contributions to $\hat{q}(k)$ become negligible for $r \gtrsim \log k$. If $r \lesssim \log k$, the previous bound implies that

$$|\hat{q}_{r+1}(k)| < e^{(\log |k|)^{C'} - c|k|}$$

and hence assumption (iii) remains essentially preserved.

Since the intervals in $\Lambda_{r+1}$ are of size $A^{-(r+1)^C}$, we may extend $\Delta_{r+1} q$ to the entire $(\lambda, \lambda')$ parameter space, with

$$\|\partial \Delta_{r+1} q\| < A^{(r+1)^C} \delta_{r+1} + A^{2(r+1)^C} \bar{\kappa}_r = \bar{\delta}_{r+1} \tag{18.38}$$

Let $q^{(r+1)} = q^{(r)} + \Delta_{r+1}q$, for which (i), (ii), and (iii) hold.

From $q^{(r+1)}$, the $Q$-equations define $\Gamma_{r+1}$ at distance from $\Gamma_r$ at most $\delta_{r+1}$. Clearly, from (18.35), also

$$\operatorname{mes}_b\left(\Gamma_{r+1} \cap (\bigcup_{I \in \Lambda_r} I' \setminus \bigcup_{I' \in \Lambda_{r+1}} I')\right) < A^{-r/4} < A^{-\frac{r+1}{5}}$$

which is (18.19). It remains to verify (b).

By (18.18), since $\operatorname{supp} F(q_r) \subset B(0, \frac{N}{10})$ with $N = A^{r+1}$ ($A$ chosen large enough),

$$\|F(q_{r+1})\| \leq \|R_{\mathbf{Z}^b \setminus B(0,N)} T T_N^{-1} R_{B(0, \frac{N}{10})}\| \, \|F(q_r)\| + \|\Delta_{r+1}q\|^2$$

$$< e^{-c\frac{N}{3}} \kappa_r + \delta_{r+1}^2$$

$\left(\text{using (c)}\right).$

Thus

$$\kappa_{r+1} < e^{-\frac{c}{3} A^{r+1}} \kappa_r + \delta_{r+1}^2 \tag{18.39}$$

Similarly,

$$\begin{aligned}
\|\partial F(q_{r+1})\| \leq{}& \|T_N^{-1}\| \, \|F(q_r)\| + \|\partial T_N^{-1}\| \, \|F(q_r)\| + \\
& \|R_{\mathbf{Z}^b \setminus B(0,N)} T T_N^{-1} R_{B(0, \frac{N}{10})}\| . \|\partial F(q_r)\| + \\
& \|\Delta_{r+1}q\| \, \|\partial \Delta_{r+1}q\| \\
<{}& A^{2(r+1)^C} \kappa_r + e^{-\frac{c}{3} A^{r+1}} \bar{\kappa}_r + \delta_{r+1} \bar{\delta}_{r+1}
\end{aligned}$$

and we may take

$$\bar{\kappa}_{r+1} = A^{2(r+1)^C} \kappa_r + e^{-\frac{c}{3} A^{r+1}} \bar{\kappa}_r + \delta_{r+1} \bar{\delta}_{r+1} \tag{18.40}$$

Summarizing (18.36), (18.38), (18.39), and (18.40),

$$\begin{cases}
\delta_{r+1} = A^{(r+1)^C} \kappa_r \\
\bar{\delta}_{r+1} = A^{2(r+1)^C} \bar{\kappa}_r + A^{(r+1)^C} \delta_{r+1} \\
\kappa_{r+1} = e^{-\frac{c}{3} A^{r+1}} \kappa_r + \delta_{r+1}^2 \\
\bar{\kappa}_{r+1} = A^{2(r+1)^C} \kappa_r + e^{-\frac{c}{3} A^{r+1}} \bar{\kappa}_r + \delta_{r+1}\bar{\delta}_{r+1}
\end{cases}$$

Start from $\kappa_0, \bar{\kappa}_0 = \varepsilon$. For $\varepsilon$ small enough, this will satisfy for $r \geq 1$

$$\begin{cases}
\delta_r < \sqrt{\varepsilon} . A^{-(\frac{4}{3})^r} & \kappa_r < \sqrt{\varepsilon} A^{-(\frac{4}{3})^{r+2}} \\
\bar{\delta}_r < \sqrt{\varepsilon} A^{-\frac{1}{2}(\frac{4}{3})^r} & \bar{\kappa}_r < \sqrt{\varepsilon} A^{-\frac{1}{2}(\frac{4}{3})^{r+2}}
\end{cases} \tag{18.41}$$

This completes the construction and proves the Theorem.

# References

[B1]  J. Bourgain. On Melnikov's persistency problem, *Math. Res. Lett.* 4 (1997), 445–458.

[B2]  J. Bourgain. Construction of quasi-periodic solutions of Hamiltonian perturbations of linear equations and applications to nonlinear PDE, *International Math. Res. Notices* 11 (1994), 475–497.

[B3]  J. Bourgain. Nonlinear Schrödinger equations, in *Hyperbolic Equations and Frequency Interactions*, IAS/Park City Math. Ser. 5, American Math. Soc., Providence, RI, 1999, pp. 3–157.

[B4]  J. Bourgain. Quasi-periodic solutions of Hamiltonian perturbations of 2D-linear Schrödinger equations, *Annals of Math.* 148 (1998), 363–439.

[C-W]  W. Craig, C. Wayne. Newton's method and periodic solutions of nonlinear wave equations, *Comm. Pure Appl. Math.* 46 (1993), 1409–1501.

[E]  L.H. Eliasson. Perturbations of stable invariant tori, *Ann. Scuola Norm. Sup. Pisa Cl. Sci.* 15(4) (1988), 115–147.

[Kuk]  S. Kuksin. Nearly integrable infinite-dimensional Hamiltonian systems, *Lecture Notes in Math.*, Springer-Verlag, Berlin, 1993, p. 1556.

[Pos]  J. Pöschel. On elliptic lower dimensional tori in Hamiltonian systems, *Math. Z.* 202 (1989), 559–608.

# Chapter Nineteen

## Application to the Construction of Quasi-Periodic Solutions of Nonlinear Schrödinger Equations

We consider next the problem of persistency of finite-dimensional tori in infinite-dimensional phase space in the context of the nonlinear Schrödinger equation (NLS) with periodic boundary conditions. More specifically, consider a nonlinear perturbation of a linear equation of the form

$$\frac{1}{i} q_t = \mathcal{L}q + \varepsilon \frac{\partial H_1}{\partial \bar{q}} \tag{19.1}$$

where $H_1 = H_1(q, \bar{q})$ is a real polynomial expression in $q, \bar{q}$ with real coefficients.

Here $q = (q_n)_{n \in \mathbf{Z}^d}$. Select a finite set of modes $n_1, \ldots, n_b \in \mathbf{Z}^d$. The operator $\mathcal{L}$ is given by a multiplier $(\mu_n)_{n \in \mathbf{Z}^d}$, where

$$\begin{cases} \mu_{n_j} = \lambda_j & (1 \le j \le b) \\ \mu_n = |n|^2 \text{ for } n \in \mathbf{Z}^d \backslash \{n_1, \ldots, n_b\} \end{cases} \tag{19.2}$$

(corresponding to the $\Delta$-operator in the Schrödinger model). Again, $\lambda = (\lambda_1, \ldots, \lambda_b)$ is a parameter taken in an interval $\Omega \subset \mathbf{R}^b$. ($A$ is a large constant depending on $H_1$ and $\varepsilon > 0$ taken small enough.)

The problem we consider is that of persistency of the unperturbed solution

$$\begin{cases} q_{n_j}(t) = a_j e^{i\lambda_j t} & (1 \le j \le b) \\ q_n(t) = 0 \text{ for } n \notin \{n_1, \ldots, n_b\} \end{cases} \tag{19.3}$$

of the linear equation ($\varepsilon = 0$).

There is the following analogue of the theorem proven in Chapter 18.

**Theorem 19.4.** *Consider (19.1) in the setting described above (with arbitrary dimension $d$). Fix $a_1, \ldots, a_b \in \mathbf{R}_+$. There is a Cantor set $\Omega_\varepsilon \subset \Omega$, mes $(\Omega \backslash \Omega_\varepsilon) \to 0$ for $\varepsilon \to 0$ and a smooth map $\lambda \to \lambda'$ on $\Omega$ s.t. for $\lambda \in \Omega_\varepsilon$, and there is a quasi-periodic solution of (19.1)*

$$q_n(t) = \sum_{k \in \mathbf{Z}^b} \widehat{q}_n(k) e^{ik.\lambda't} \qquad (n \in \mathbf{Z}^d)$$

*satisfying*

$$\widehat{q}_{n_j}(e_j) = a_j \qquad (1 \le j \le b)$$
$$|\widehat{q}_n(k)| \lesssim e^{-c(|n|+|k|)}$$

*and*

$$\sum_{(n,k) \notin \mathcal{R}} |\widehat{q}_n(k)| < \sqrt{\varepsilon}$$

*where*

$$\mathcal{R} = \{(n_j, e_j) | j = 1, \ldots, b\}$$

This result is proven in [B2] for $d = 1$ and in [B4] for $d = 2$ (see references from Chapter 4). The proof given here is along the lines of the preceding chapter and extends to general dimension. We will follow the same method as explained in Chapter 18. Identifying $q$ and $\hat{q} = \hat{q}(n, k) = \hat{q}_n(k)$, rewrite (19.1) as

$$(k.\lambda' - \mu_n)\hat{q}(n, k) - \varepsilon \widehat{\frac{\partial H_1}{\partial \bar{q}}}(n, k) = 0 \qquad (n \in \mathbf{Z}^d, k \in \mathbf{Z}^b) \qquad (19.5)$$

where we specify the Fourier coefficients

$$\widehat{q_{n_j}}(e_j) = a_j \qquad (1 \le j \le b)$$

To solve (19.5), make again a decomposition in $P$- and $Q$-equations. The $P$-equations are given by

$$\begin{cases} (k.\lambda' - \mu_n)u(n, k) - \varepsilon \widehat{\frac{\partial H_1}{\partial \bar{q}}}(n, k) = 0 \text{ where } (n, k) \notin \mathcal{R} = \{(n_j, e_j) | j = 1, \ldots, b\} \\ (-k.\lambda' - \mu_n)v(n, k) - \varepsilon \widehat{\frac{\partial H_1}{\partial q}}(n, k) = 0 \text{ where } (n, k) \notin -\mathcal{R} \end{cases}$$

The linearized operator $T = T_q$ is given by

$$T_q = D - \varepsilon S_q$$

where $D$ is diagonal with index set $(\pm, n, k)$

$$D_{\pm, n, k} = \pm k.\lambda' - \mu_n$$

and

$$S_q = \begin{pmatrix} S_{\frac{\partial^2 H_1}{\partial \bar{q} \partial q}} & S_{\frac{\partial^2 H_1}{\partial \bar{q}^2}} \\ S_{\frac{\partial^2 H_1}{\partial q^2}} & S_{\frac{\partial^2 H_1}{\partial q \partial \bar{q}}} \end{pmatrix}$$

and where $S_\phi$ denotes the Toeplitz operator with symbol $\phi$.

Letting $q = q^{(r)}$, $N = A^r$, the main issue is again to establish bounds

$$\|T_N^{-1}\| < A^{r^C} \qquad (19.6)$$

and (denoting $(n, k) = \xi$ for simplicity)

$$|T_N^{-1}(\xi, \xi')| < e^{-c|\xi - \xi'|} \text{ for } |\xi - \xi'| > r^C \qquad (19.6')$$

Here $T = T_{q^{(r)}}$ and $T_N$ refers to the restriction of $T$ in both indices $k$ and $n$, thus

$$T_N = R_{\substack{|k| < N \\ |n| < N}} T R_{\substack{|k| < N \\ |n| < N}}$$

For (19.6) and (19.6') to hold, a restriction $(\lambda, \lambda') \in \bigcup_{I \in \Lambda_r} I$ will be made, as in the previous chapter. But in the present situation this restriction also will involve some additional issues. Once this fact is obtained, the argument to prove Theorem 19.4 may be carried out almost exactly as in the preceding chapter.

The inversion of $T_N$ will be more complicated here, and the analogue of Lemma 18.27 is more delicate. We introduce again the matrix $T^\sigma$, depending on the extra parameter $\sigma$ (see (18.24))

$$T^\sigma = D^\sigma + \varepsilon S_{q^{(r_0)}(\lambda, \lambda')} \qquad (19.7)$$

with

$$D^\sigma_{\pm,n,k} = \pm(k.\lambda' + \sigma) - \mu_n \qquad (19.8)$$

For $M_0 \in \mathbf{Z}_+$ and $Q$ a $d$-dimensional interval in $\mathbf{Z}^d$, denote $T^\sigma_{M_0,Q}$ the restriction to $k \in \mathbf{Z}^b$, $|k| < M_0$ and $n \in Q$.

Observe that from (19.8), for $n \notin \{n_1, \ldots, n_b\}$,

$$|D^\sigma_{\pm,n,k}| > \big| |n|^2 - |k.\lambda' + \sigma| \big|$$

and hence, if $|D^\sigma_{\pm,n,k}| < 1, |D^\sigma_{\pm,n',k'}| < 1$, then

$$\big| |n|^2 - |n'|^2 \big| < C|k - k'| + 2 \qquad (19.9)$$

We will use the following arithmetical lemma, a proof for which will be given at the end of this chapter.

**Lemma 19.10.** *Fix any large number $B$. There is a partition $\{\pi_\alpha\}$ of $\mathbf{Z}^d$ satisfying the properties*

$$\operatorname{diam} \pi_\alpha < B^{C_0} \qquad (19.11)$$

*and*

$$|n - n'| + \big| |n|^2 - |n'|^2 \big| > B \ \text{if} \ n \in \pi_\alpha, n' \in \pi_{\alpha'}, \alpha \neq \alpha' \qquad (19.12)$$

*(where $C_0 = C(d)$).*

To treat $(T^\sigma_{M_0,Q})^{-1}$, consider first the (easier) case where $Q$ is an $M_0$-interval in $\mathbf{Z}^d$ satisfying

$$\min_{n \in Q} |n| > M_0$$

In this situation, we prove

**Lemma 19.13.** *Let $T = T_q$ with $q = q(\lambda, \lambda')$ be holomorphic in $\lambda, \lambda'$ on an interval $I \subset \mathbf{R}^{2b}$ and $\|\partial q\| < C$ ($\partial = \partial_\lambda$ or $\partial_{\lambda'}$). We also assume $\lambda'$ satisfying a DC*

$$\|k.\lambda'\| > M_0^{-C} \ \text{for} \ k \in \mathbf{Z}^b, 0 < |k| < M_0$$

*Consider $T^\sigma_{M_0,Q}$ with $\min_{n \in Q} |n| > M_0$. There is a system of Lipschitz functions $\sigma_s = \sigma_s(\lambda, \lambda'), s < e^{(\log M_0)^{C_1}}$ (depending on $Q$) such that*

$$\|\sigma_s\|_{\mathrm{Lip}} \leq CM_0 \qquad (19.14)$$

*and for $\lambda, \lambda' \in I$*

$$\|(T^\sigma_{M_0,Q})^{-1}\| < e^{M_0^{1-}} \qquad (19.15)$$

*and*

$$|(T^\sigma_{M_0,Q})^{-1}(\xi, \xi')| < e^{-c|\xi - \xi'|} \ \text{for} \ |\xi - \xi'| > \frac{M_0}{10} \qquad (19.15')$$

*provided*

$$\min_s |\sigma - \sigma_s(\lambda, \lambda')| > e^{-M_0^{c_2}} \qquad (19.16)$$

*($C_1, c_2$ are some constants to which we will refer later on).*

Lemma 19.13 is proven multiscale in $M_0$ and is mainly based on Lemma 19.10, first-order eigenvalue variation, and the resolvent identity.

Assume Lemma 19.13 valid at scale $M_0$, and let

$$M_1 = e^{M_0^{\frac{1}{2}c_2}} \tag{19.17}$$

and $Q \subset \mathbf{Z}^b$ an $M_1$-interval s.t. $\min_{n \in Q} |n| > M_1$.

Let $Q_0 = [-M_0, M_0]^d$, and consider the matrices $T_{M_0, n_0 + Q_o}^{\sigma + k.\lambda'}$ where $|k| < M_1, n_0 \in Q$

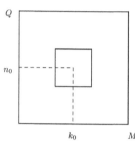

Choose $B = M_1^\rho$ ($0 < \rho < 1$ to be specified) in Lemma 19.10 and let $\{\pi_\alpha\}$ be the corresponding partition of $Q_1$.

Call a site $(\pm, n, k)$ singular if $|D_{\pm, n, k}^\sigma| < 1$. Assume $(\pm, n, k)$ and $(\pm, n', k')$ singular with $n \in \pi_\alpha$, $n' \in \pi_{\alpha'}$, $\alpha \neq \alpha'$. It then follows from (19.9) and (19.12) that

$$|n - n'| + |k - k'| > cM_1^\rho \tag{19.18}$$

Therefore, if we denote

$$\mathcal{S} = \{(n, k) \in Q \times [-M_1, M_1]^b | (+, n, k) \text{ or } (-, n, k) \text{ is singular}\} \tag{19.19}$$

the following separation property holds

$$\text{dist}\left(\mathcal{S} \cap (\mathbf{Z}^b \times \pi_\alpha), \mathcal{S} \cap (\mathbf{Z}^b \times \pi_{\alpha'})\right) \gtrsim M_1^\rho \text{ for } \alpha \neq \alpha' \tag{19.20}$$

Fix $n_0 \in \mathbf{Z}^d$. From the induction hypothesis, $T_{M_0, n_0 + Q_0}^{\sigma + k.\lambda'}$ will satisfy (19.15), (19.15') unless

$$\min_s |\sigma + k.\lambda' - \sigma_s(\lambda, \lambda')| < e^{-M_0^{c_2}}$$

Since $\|k.\lambda'\| > M_1^{-C}$ for $k \in \mathbf{Z}^b, 0 < |k| < M_1$, it follows from (19.17) that, for fixed $s$, $|\sigma + k.\lambda' - \sigma_s(\lambda, \lambda')| < e^{-M_0^{c_2}}$ holds for at most one value of $k$. Thus, for fixed $\alpha$, there are clearly at most (recalling (19.11))

$$M_1^{\rho C_0 \cdot d} \times e^{(\log M_0)^{C_1}} < M_1^{\frac{1}{100}} \tag{19.21}$$

sites $(k, n) \in [-M_1, M_1]^b \times \pi_\alpha$ for which $(T_{M_0, n+Q_0}^{k.\lambda'+\sigma})^{-1}$ fails (19.15) and (19.15'). For these $(n, k)$, necessarily $((k + [-M_0, M_0]^b) \times (n + Q_0)) \cap \mathcal{S} \neq \phi$. Partitioning $\Omega = [-M_1, M_1]^b \times Q$ in intervals of size $M_0$, the preceding permits us to perform a decomposition

$$\Omega = \Omega_0 \cup \Omega_1 \tag{19.22}$$

with

$$\Omega_1 = \bigcup_\beta \Omega_{1,\beta} \tag{19.23}$$

where $\Omega_0, \Omega_{1\beta}$ are unions of $M_0$-interval

$$\operatorname{diam} \Omega_{1,\beta} < M_1^{\frac{1}{100}+\rho} + M_1^{\rho C_0} \tag{19.24}$$

$$\operatorname{dist}(\Omega_{1,\beta}, \Omega_{1,\beta'}) > M_1^\rho \text{ for } \beta \neq \beta' \tag{19.25}$$

and $(T_{\Omega_0}^\sigma)^{-1}$ satisfies (as a consequence of the resolvent identity)

$$\|(T_{\Omega_0}^\sigma)^{-1}\| < e^{M_0^{-1}} \tag{19.26}$$

and

$$|(T_{\Omega_0}^\sigma)^{-1}(\xi,\xi')| < e^{-c|\xi-\xi'|} \text{ for } |\xi-\xi'| > 10M_0 \tag{19.26'}$$

Notice here that clearly (19.26) and (19.26') allow an $e^{-M_0}$-perturbation of $T^\sigma$ and hence an $0(M_1^{-1}e^{-M_0})$-perturbation of $(\lambda, \lambda', \sigma)$. The decomposition (19.22) and (19.23) thus may be used on an $M_1^{-1}e^{-M_0}$-neighborhood of an initial parameter choice $(\lambda, \lambda', \sigma)$. Denote $\widetilde\Omega_{1,\beta}$ an $M_1^{\frac{3}{4}\rho}$-neighborhood of $\Omega_{1,\beta}$. If we ensure that

$$\|(T_{\widetilde\Omega_{1,\beta}}^\sigma)^{-1}\| < e^{M_1^{\rho/2}} \tag{19.27}$$

for all $\beta$, (19.22), (19.23), (19.25), (19.26), (19.26'), (19.27), and the resolvent identity imply that

$$\|(T_\Omega^\sigma)^{-1}\| < e^{3M_1^{\rho/2}} < e^{M_1^{1-}} \tag{19.28}$$

and

$$|(T_\Omega^\sigma)^{-1}(\xi,\xi')| < e^{-c|\xi-\xi'|} \text{ for } |\xi-\xi'| > \frac{M}{10} \tag{19.28'}$$

Consider condition (19.27). Recall that for $(k,n) \in \Omega$,

$$D_{\pm,n,k}^\sigma = \pm(k.\lambda' + \sigma) - |n|^2 = \pm\sigma - |n|^2 + 0(M_1)$$

with $n \in Q$; hence $|n| > M_1$. Assume $\sigma > 0$, thus $|-\sigma - |n|^2| > M_1^2$. Hence

$$\|(R_-T_{\widetilde\Omega_{1,\beta}}^\sigma R_-)^{-1}\| \lesssim M_1^{-2}$$

and $(T_{\widetilde\Omega_{1,\beta}}^\sigma)^{-1}$ is controlled by the inverse of the self-adjoint matrix

$$R_+T_{\widetilde\Omega_{1,\beta}}^\sigma R_+ - \varepsilon^2 R_+S_qR_-(R_-T_{\widetilde\Omega_{1,\beta}}^\sigma R_-)^{-1}R_-S_qR_+ \tag{19.29}$$

$$= \sigma + R_+T_{\widetilde\Omega_{1,\beta}}R_+ - \varepsilon^2 R_+S_qR_-(R_-T_{\widetilde\Omega_{1,\beta}}^\sigma R_-)^{-1}R_-S_qR_+$$

$$= \sigma + T'(\lambda, \lambda', \sigma) \tag{19.30}$$

Clearly, $\|\partial_\lambda T'\| \leq \|\partial_\lambda q\| < C$, and similarly, $\|\partial_{\lambda'} T'\| < M_1$. Also, $\|\partial_\sigma T'\| < C\varepsilon^2\|(R_-T_{\widetilde\Omega_{1,\beta}}^\sigma R_-)^{-1}\|^2 < M_1^{-4}$.

Let $\{E_i(T')\}$ be the ascending ordering of the eigenvalues of $T'$. Then $E_i$ are continuous functions of the parameters $\lambda, \lambda', \sigma$ and (from the analyticity assumption on $q = q(\lambda, \lambda'))$ piecewise holomorphic in each (1-dim) parameter component separately. From the preceding and first-order eigenvalue variation, it follows that $E_i(\lambda, \lambda', \sigma)$ is Lipschitz and

$$\|E_i\|_{\text{Lip}(\lambda)} < C, \|E_i\|_{\text{Lip}(\lambda')} < M_1, \|E_i\|_{\text{Lip}(\sigma)} < M_1^{-4} \tag{19.31}$$

From (19.30), $\{\sigma + E_i(\lambda, \lambda', \sigma)\}$ is a parametrization of Spec (19.29). Fix $i$. By (19.31), the equation

$$\sigma + E_i(\lambda, \lambda', \sigma) = 0$$

defines a function

$$\sigma = \sigma_i(\lambda, \lambda')$$

Thus

$$\sigma_i(\lambda, \lambda') + E_i(\lambda, \lambda', \sigma_i(\lambda, \lambda')) = 0$$

and again by (19.31),

$$|\sigma_i(\lambda, \lambda') - \sigma_i(\bar{\lambda}, \bar{\lambda}')| \leq M_1^{-4}|\sigma_i(\lambda, \lambda') - \sigma_i(\bar{\lambda}, \bar{\lambda}')| + C|\lambda - \bar{\lambda}| + M_1|\lambda' - \bar{\lambda}'|$$

implying

$$\|\sigma_i\|_{\text{Lip}(\lambda)} \leq C, \|\sigma_i\|_{\text{Lip}(\lambda')} < M_1 \tag{19.32}$$

Moreover,

$$|\sigma + E_i(\lambda, \lambda', \sigma)| = |\sigma - \sigma_i(\lambda, \lambda') + E_i(\lambda, \lambda', \sigma) - E_i(\lambda, \lambda', \sigma_i(\lambda, \lambda'))|$$
$$= |\sigma - \sigma_i(\lambda, \lambda')|(1 + 0(M_1^{-4}))$$

Consequently,

$$\text{dist}\,(\text{Spec}\,(19.29), 0) \sim \min_i |\sigma - \sigma_i(\lambda, \lambda')|$$

Hence

$$\|(T^\sigma_{\bar{\Omega}_{1,\beta}})^{-1}\| \leq \max_i |\sigma - \sigma_i(\lambda, \lambda')|^{-1} \tag{19.33}$$

Collecting the functions $\{\sigma_i\}$ over all $\beta$, we thus obtain at most $M_1^{b+d}$ Lipschitz functions $\{\sigma_i = \sigma_i(\lambda, \lambda')\}$ such that

$$\|\sigma_i\|_{\text{Lip}} \lesssim M_1 \tag{19.34}$$

and

$$\max_\beta \|(T^\sigma_{\bar{\Omega}_{1,\beta}})^{-1}\| \leq \max_i |\sigma - \sigma_i(\lambda, \lambda')|^{-1} \tag{19.35}$$

If $(19.35) < e^{M_1^{\rho/2}}$, (19.27) will hold, hence (19.28) and (19.28′).

The preceding construction of the $\{\sigma_i\}$-functions has been carried out after restriction of the $(\lambda, \lambda', \sigma)$-parameters to an $M_1^{-1}e^{-M_0}$-neighborhood of an initial choice. As pointed out earlier, one may indeed keep the same decomposition (19.22) and (19.23) on such a neighborhood. The number of these parameter neighborhoods clearly may be bounded by

$$M_1^{d+1}(M_1 e^{M_0})^{2b+1}$$

and the total number of $\{\sigma_i\}$ functions therefore is at most

$$M_1^{b+d} M_1^{d+1} (M_1 e^{M_0})^{2b+1} < e^{2(b+1)M_0} < e^{2(b+1)(\log M_1)^{2/c_2}}$$

by (19.17).

From the preceding, if

$$\min_i |\sigma - \sigma_i(\lambda, \lambda')| < e^{-M_1^{\rho/2}}$$

then (19.28) and (19.28′) hold.

Recalling condition (19.21) on $\rho$, i.e., $M_1^{\rho C_0 d} e^{C(\log\log M_1)^{C_1}} < M_1^{\frac{1}{100}}$, we take

$$\rho = \frac{1}{200 C_0 . d} \tag{19.36}$$

This proves Lemma 19.19 with $c_2 = \frac{\rho}{2}, C_1 = 3/c_2$.

Next, we will prove an estimate on $(T_{M_0,Q}^\sigma)^{-1}$ with $Q = [-10M_0, 10M_0]^d$. To establish this estimate, some further restrictions in $(\lambda, \lambda')$-parameter space will be needed. In view of this, we make the following definition:

**Definition.** *Say that $\mathcal{A} \subset [0,1]^{2b}$ has sectional measure at most $\varepsilon$, $\operatorname{mes}_{sec}\mathcal{A} < \varepsilon$, if the following holds:*
*Let $\varphi : [0,1]^b \to [0,1]^b$ be $C^1$ with $\|\nabla\varphi\| < 10^{-2}$. Then*

$$\operatorname{mes}\left[\lambda \in [0,1]^b \,\middle|\, (\lambda, \lambda + \varphi(\lambda)) \in \mathcal{A}\right] < \varepsilon \tag{19.37}$$

As part of a multiscale reasoning, we next prove

**Lemma 19.38.** *Let $T = T_q$ with $q = q(\lambda, \lambda')$ be holomorphic in $\lambda, \lambda'$ on an interval $I \subset [0,1]^{2b}$ and $\|\partial q\| < C$. Let $M_0 \in \mathbf{Z}_+$ be a large integer*

$$M_0 < M_1 < \exp\exp(\log M_0)^{\frac{1}{10}} \tag{19.39}$$

*Assume $\lambda'$ satisfies a DC*

$$\|k.\lambda'\| > M_1^{-C} \text{ for } |k| < M_1$$

*Assume that $T_M^\sigma = T_{M,Q=[-10M,10M]^d}^\sigma$ satisfies for $M < M_0$ and $(\lambda, \lambda') \in I$ the following property*

$$\|(T_M^\sigma)^{-1}\| < e^{M^{1-}} \tag{19.40}$$

*and*

$$|(T_M^\sigma)^{-1}(\xi, \xi')| < e^{-c|\xi-\xi'|} \text{ for } |\xi - \xi'| > \frac{1}{10}M \tag{19.40′}$$

*for all $\sigma$ outside a set of measure*

$$< e^{-M^{c_3}} \tag{19.41}$$

*Then there is a subset $\mathcal{A}$ of $I$ obtained as a union of intervals of size $[\exp\exp(\log\log M_1)^3]^{-1}$, such that*

$$\operatorname{mes}_{sec}\mathcal{A} < [\exp\exp(\log\log M_1)^2]^{-1} \tag{19.42}$$

*and if $(\lambda, \lambda') \in I \backslash \mathcal{A}$*

$$\|(T^\sigma_{M_1})^{-1}\| < e^{M_1^{\frac{1}{10}}} \tag{19.43}$$

*for $\sigma$ outside a set (depending on $\lambda, \lambda'$) of measure*

$$< e^{-M_1^{c_4}} \qquad (c_4 = 10^{-7}d^{-1}c_2) \tag{19.44}$$

**Proof.** Let

$$M = \exp(\log \log M_1)^3 < \exp(\log M_0)^{3/10} \tag{19.45}$$

Start by paving $[-M_1, M_1]^b \times [-10M_1, 10M_1]^d$ by translates of $[-M, M]^b \times [-M, M]^d$. Thus we need to control $(T^{\sigma+k.\lambda'}_{M,Q})^{-1}$ with $|k| < M_1$ and $Q \subset [-10M_1, 10M_1]^d$ an $M$-interval. If $\min_{n \in Q} |n| > M$, use Lemma 19.13 applied at scale $M$. For $(T^{\sigma+k.\lambda'}_{M,Q})^{-1}$ to satisfy (19.15) and (19.15'), we need thus to satisfy

$$\min_s |\sigma + k.\lambda' - \sigma_s(\lambda, \lambda')| > e^{-M^{c_2}} \tag{19.46}$$

where $\{\sigma_s\}$ is the system of Lipschitz functions introduced in Lemma 19.13. These depend on $Q$. Collecting then over all $M$-intervals $Q$ in $[-10M_1, 10M_1]^d \backslash [|n| < M]$, their number is at most

$$(20M_1)^d . e^{(\log M)^{C_1}} < M_1^{d+1} \tag{19.47}$$

and we still denote $\{\sigma_s\}$ this collected system. We introduce $\mathcal{A} \subset I$ s.t. for $(\lambda, \lambda') \in I \backslash \mathcal{A}$ and any $k \in \mathbf{Z}^b$, $M^{1+} < |k| \le M_1$,

$$\min_{s,s'} |k\lambda' - \sigma_s(\lambda, \lambda') + \sigma_{s'}(\lambda, \lambda')| > 2e^{-M^{c_2}} \tag{19.48}$$

If (19.48) holds, it will follow that if $\min_{n \in Q \cup Q'} |n| > M$ and $(T^{\sigma+k.\lambda'}_{M,Q})^{-1}$, $(T^{\sigma+k'.\lambda}_{M,Q'})^{-1}$ both fail (19.15) and (19.15'), then $|k - k'| < M^{1+}$. Thus, if $\min_{n \in Q} |n| > M$ and $(T^{\sigma+k.\lambda'}_{M,Q})^{-1}$ fails (19.15) and (19.15'), then $k$ is within an $M^{1+}$-neighborhood of a single $\tilde{k} \in [M_1, M_1]^b$. Since necessarily

$$\min_{|k'-k|<M, n \in Q} |\pm (k'.\lambda' + \sigma) - |n|^2| < 1$$

it follows also that

$$\min_{n \in Q} | \, |n|^2 - |\tilde{k}.\lambda' + \sigma| \, | < M + M^{1+} = M^{1+}$$

Therefore, there are $\tilde{k} \in [-M_1, M_1]^b$ and $R > 0$, s.t.

$$|k - \tilde{k}| + \min_{n \in Q} | \, |n|^2 - R^2| < M^{1+} \tag{19.49}$$

Unless (19.49) holds, $(T^{\sigma+k.\lambda'}_{M,Q})^{-1}$ will satisfy (19.15) and (19.15').

Returning to (19.48), there are by (19.47) at most $M_1^b . M_1^{2(d+1)} = M_1^{b+2d+2}$ conditions involved. We claim that if $|k| > M^{1+}$ and $\varphi : [0,1]^b \to [0,1]^b$ satisfies $|\nabla\varphi| < 10^{-2}$, then

$$\text{mes}\,[\lambda| \,|k.(\lambda + \varphi(\lambda)) - \sigma_s(\lambda, \lambda + \varphi(\lambda)) + \sigma_{s'}(\lambda, \lambda + \varphi(\lambda))| \le 2e^{-M^{c_2}}] < e^{-M^{c_2}} \tag{19.50}$$

This will imply that

$$\text{mes}\,\text{sec}\,\mathcal{A} < M_1^{b+2d+2}e^{-M^{c_2}} < [\exp\exp(\log\log M_1)^2]^{-1} \tag{19.51}$$

Clearly, $\mathcal{A}$ may moreover be taken to be a union of intervals of size $e^{-M}$.

To check (19.50), assume, for instance, that $k_1 > M^{1+}$. Considering $k.(\lambda + \varphi(\lambda)) - \sigma_s(\lambda, \lambda + \varphi(\lambda)) + \sigma_{s'}(\lambda, \lambda + \varphi(\lambda))$ as a function in $\lambda_1$ with other variables fixed, we have

$$\partial_{\lambda_1}[k.(\lambda + \varphi(\lambda))] > \frac{99}{100}k_1 > M^{1+}$$

while by (19.14)

$$\|(\sigma_s - \sigma_{s'})(\lambda, \lambda + \varphi(\lambda))\|_{\text{Lip}(\lambda_1)} \lesssim M$$

This clearly implies (19.50).

Consider next $T_{M^2}^{\sigma+k.\lambda'}$, $|k| < M_1$. Use the induction hypothesis and in particular (19.41) at scale $M^2 < M_0$. Since conditions (19.40) and (19.40') are semialgebraic in $\sigma$ of degree $< (CM)^{b+d}$ and $\lambda'$ assumed diophantine as specified in Lemma 19.38, it follows that $(T_{M^2}^{\sigma+k.\lambda'})^{-1}$ fails (19.40) and (19.40') for at most $M^C$ values of $k$ (we use here (19.45)).

Partitioning $\Omega = [-M_1, M_1]^b \times [-10M_1, 10M_1]^d$ in $M$-intervals, the preceding and the resolvent-identity provide a decomposition

$$\Omega = \Omega_0 \cup \Omega_1 \tag{19.52}$$

where $\Omega_0, \Omega_1$ are unions of $M$-intervals

$$\#\pi_1(\Omega_1) < M^C + M^{1+} \tag{19.53}$$

($\pi_1 =$ projection in $k$-variable) and

$$(k, n) \in \Omega_1 \Rightarrow |n| < M^2 \text{ or } \big| |n| - R \big| < \frac{M^{1+}}{R} \tag{19.54}$$

and such that

$$\|(T_{\Omega_0}^\sigma)^{-1}\| < e^{M^2} \tag{19.55}$$

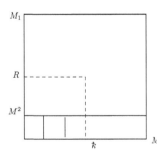

Our purpose is to apply Proposition 14.1 with $A(\sigma) = T_{M_1}^\sigma$. Thus $B_1 < C$. Take in (14.5) $\Lambda = \Omega_1$ satisfying by (19.53) and (19.54)

$$\#\Lambda < M^C.R^{d-2+} \tag{19.56}$$

and $B_2 = e^{M^2}$ by (19.55).

To obtain condition (14.6), simply apply Lemma 19.13, the induction hypothesis at scale $M_0 < M_1$, together with the resolvent identity. Thus

$$\text{mes}\left[\sigma \mid \|(T_{M_1}^\sigma)^{-1}\| > B_3 = e^{M_0}\right] < M_1^{b+d} e^{(\log M_0)^{C_1}} e^{-M_o^{c_2}} + M_1^b . e^{-M_0^{c_3}} \ll \frac{1}{B_2} \tag{19.57}$$

We make the following distinction (not necessary for $d \leq 2$)

**Case I:** $R \leq M_1^{\frac{1}{100d}}$

**Case II:** $R > M_1^{\frac{1}{100d}}$

**Case I.** By (19.56), $\#\Lambda < M_1^{\frac{1}{100}}$. Proposition 14.1 then permits us to conclude that

$$\text{mes}\left[\sigma \mid \|(T_{M_1}^\sigma)^{-1}\| > e^{M_1^{\frac{1}{10}}}\right] < \exp\left\{ -\frac{cM_1^{\frac{1}{10}}}{(\#\Lambda)\log(M_1+B_2+B_3)} \right\} \tag{19.58}$$

$$< e^{-cM_1^{\frac{1}{10} - \frac{1}{100}}} M_0^{-1} < e^{-M_1^{\frac{1}{15}}}$$

**Case II.** Then apply Proposition 14.1 to $A(\sigma) = T_{M_1,[|n|<M_1^{10^{-2}d-1}]}^\sigma$, for which (19.53) and (19.54) imply $\#\Lambda < M^C$. It follows from Proposition 14.1 that

$$\text{mes}\left[\sigma \mid \|(T_{M_1,[|n|<M_1^{10^{-2}d-1}]}^\sigma)^{-1}\| > e^{M_1^{10^{-6}d-1}}\right]$$

$$< \exp\{-cM_1^{10^{-6}d-1} M^{-C} M_0^{-1}\} \tag{19.59}$$

$$< e^{-M_1^{10^{-7}d-1}}$$

To control $(T_{M_1,[|n|>M_1^{10^{-3}d-1}]}^\sigma)^{-1}$, we simply use Lemma 19.13. More precisely, define

$$M_2 = M_1^{10^{-4}d-1} \tag{19.60}$$

and cover $[|k| < M_1] \times [M_1^{10^{-3}d-1} < |n| < 10M_1]$ by $(b+d)$-dim intervals $P$ of size $M_2$. From Lemma 19.13, applied at scale $M_2$,

$$\|(T_P^\sigma)^{-1}\| < e^{M_2^{1-}} \tag{19.61}$$

$$|(T_P^\sigma)^{-1}(\xi,\xi')| < e^{-c|\xi-\xi'|} \text{ for } \xi, \xi' \in P, |\xi - \xi'| > \frac{M_2}{10} \tag{19.61'}$$

for $\sigma$ outside a set of measure at most

$$M_1^{b+d} . e^{(\log M_2)^{C_1}} . e^{-M_2^{c_2}} < e^{-M_1^{c_2 10^{-5}d-1}} \tag{19.62}$$

From (19.59), (19.60), (19.61), (19.61'), and another application of the resolvent identity, we conclude that for $\sigma$ outside a set of measure at most

$$e^{-M_1^{10^{-7}d-1}} + e^{-M_1^{c_2 10^{-5}d-1}} < e^{-M_1^{c_2 10^{-7}d-1}} \tag{19.63}$$

one has that

$$\|(T_{M_1}^\sigma)^{-1}\| < e^{M_1^{10^{-4}d-1}} < e^{M_1^{\frac{1}{10}}} \tag{19.64}$$

Together with (19.58), this completes the proof.

**Lemma 19.65.** *In the statement of Lemma 19.38 we have in addition to (19.43) also the off-diagonal decay estimate*

$$|(T^{\sigma}_{M_1})^{-1}(\xi, \xi')| < e^{-c|\xi-\xi'|} \text{ for } |\xi - \xi'| > \frac{1}{10}M_1 \qquad (19.66)$$

*and $\sigma$ outside a set of measure $< e^{-M_1^{c_5}}$, $c_5 = c_2 10^{-12} d^{-1}$.*

**Proof.** Define again $M_2 = M_1^{10^{-4}d^{-1}}$. Consider first $M_2$ intervals $Q \subset [-10M_1, 10M_1]^d$ s.t. $\min_{n \in Q} |n| > M_2$. Lemma 19.13 implies that

$$\|(T^{\sigma+k.\lambda'}_{M_2,Q})^{-1}\| < e^{M_2^{1-}} \qquad (19.67)$$

and

$$|(T^{\sigma+k.\lambda'}_{M_2,Q})^{-1}(\xi, \xi')| < e^{-c|\xi-\xi'|} \text{ for } |\xi - \xi'| > \frac{1}{10}M_2 \qquad (19.67')$$

for all $k \in [-M_1, M_1]^b$ and $\sigma$ outside a set of measure at most

$$M_1^{b+d} . e^{(\log M_2)^{C_1}} . e^{-M_2^{c_2}} < e^{-\frac{1}{2}M_2^{c_2}} < e^{-M_1^{10^{-5}d^{-1}c_2}} \qquad (19.68)$$

It remains to consider the lower region $[-M_1, M_1]^b \times [-100M_2, 100M_2]^d$.

Take $M$ as in the proof of Lemma 19.38. Consider a paving of $[-M_1, M_1]^b \times [-100M_2, 100M_2]^d$ with intervals $P = [k - M, k + M]^b \times [-10M, 10M]^d$ and intervals $P' = [k - M, k + M]^b \times Q$ with $Q$ an $M$-interval in $[-100M_2, 100M_2]^d$ s.t. $\min_{n \in Q} |n| > M$.

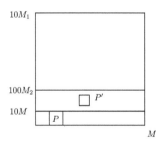

For the $P$-intervals, the assumption in Lemma 19.38 implies again that $(T^{\sigma}_P)^{-1}$ will satisfy (19.40) and (19.40') except for $M^C$-many values of $k$. For the $P'$ intervals, Lemma 19.13 applies again, and the number of "bad" intervals there is at most

$$e^{(\log M)^{C_1}} (100M_2)^d < M_1^{2.10^{-4}}$$

Thus the total number of "bad" $P$ and $P'$ intervals in $[-M_1, M_1]^b \times [-100M_2, 100M_2]^d$ is at most $M_1^{2.10^{-4}}$ (for a fixed $\sigma$).

Define

$$M_3 = M_1^{10^{-3}}$$

It is clear from the preceding that for any $k \in [-M_1, M_1]^b$, there is some $\tilde{k} \in [-M_1, M_1]^b$ and some $M_3 < M' < 2M_3$ s.t.

$$k \in \tilde{k} + [-M', M']^b \subset [-M_1, M_1]^b$$

and

$$\left(\mathring{k} + ([-M', M']^b \setminus [-M' + M_3^{1/2}, M' - M_3^{1/2}]^b)\right)$$
$$\cap \left(\bigcup_{P \, bad} \pi_1(P) \cup \bigcup_{P' \, bad} \pi_1(P')\right) = \phi \tag{19.69}$$

Lemma 19.38 permits us to ensure for $(\lambda, \lambda') \in I \setminus \mathcal{A}$ that

$$\|(T_{M'}^{\sigma + \mathring{k}\lambda'})^{-1}\| < e^{(M')^{\frac{1}{10}}} < e^{M_3^{\frac{1}{9}}} \text{ for all } \mathring{k} \in [-M_1, M_1]^b \text{ and } M_3 < M' < 2M_3 \tag{19.70}$$

except for $\sigma$ in a set of measure

$$< M_3 M_1^b e^{-M_3^{c_4}} < e^{-M_1^{10^{-11} d^{-1} c_2}} \tag{19.71}$$

Thus, excluding a $\sigma$-set of measure

$$e^{-M_1^{10^{-5} d^{-1} c_2}} + e^{-M_1^{10^{-11} d^{-1} c_2}} < e^{-M_1^{c_5}} \tag{19.72}$$

(recalling also (19.68)), properties (19.67), (19.67′), (19.69), and (19.70), together with the resolvent identity, imply (19.66).

This proves Lemma 19.65.

**Remarks.**

**1.** Lemma 19.65 permits us to recover at scale $M_1$ the assumptions form Lemma 19.38 at scale $M \leq M_0$, with $c_3 = c_5 = 10^{-12} d^{-1} c_2$. There is, however, an additional restriction $(\lambda, \lambda') \in I \setminus \mathcal{A}$, where

$$\text{mes}_{\text{sec}} \mathcal{A} < [\exp \exp (\log \log M_1)^2]^{-1}$$

**2.** A comment on the assumption for $q = q(\lambda, \lambda')$ to be holomorphic in $\lambda, \lambda'$ on $I \subset [0, 1]^{2b}$. Recall that $q = q^{(r)}$ is obtained along the Newton iteration scheme and satisfies for $r_0 < r$

$$\|q^{(r)} - q^{(r_0)}\| < \delta_{r_0} \text{ where } \log \log \frac{1}{\delta_{r_0}} \sim r_0 \tag{19.73}$$

Clearly, one may replace $q$ by $q^{(r_0)}$ provided

$$\delta_{r_0} < e^{-M_1} \text{ hence } r_0 \sim \log M_1 \tag{19.74}$$

This approximative solution $q^{(r_0)} = q^{(r_0)}(\lambda, \lambda')$ is given by a rational function with bounded derivatives in $(\lambda, \lambda')$ for $(\lambda, \lambda') \in \bigcup_{I \in \Lambda_{r_0}} I$, consisting of intervals $I$ of size $A^{-r_0^C}$ in $[0, 1]^{2b}$. If Lemmas 19.38 and 19.65 are applied with $I$ one of those intervals, the set $\mathcal{A}$ removed from $I$ thus satisfies

$$\text{mes}_{\text{sec}} \mathcal{A} < [\exp \exp (\log r_0)^2]^{-1} \ll A^{-r_0^C} \tag{19.75}$$

We return now to Chapter 18 and the inductive construction. In order to be able to estimate $T_N^{-1}, N = A^r$, we add the following hypothesis:

(†) *Let* $\exp(\log \log M)^3 \leq r_1 \leq r$. *Then for all* $(\lambda, \lambda') \in I \in \Lambda_{r_1}$, $T_{M,Q}^{\sigma}$, *with* $Q = [-10M, 10M]^d$, *satisfies*

$$\|(T_{M,Q}^{\sigma})^{-1}\| < e^{M^{1-}} \tag{19.76}$$

*and*

$$|(T_{M,Q}^{\sigma})^{-1}(\xi, \xi')| < e^{-c|\xi - \xi'|} \text{ for } |\xi - \xi'| > \frac{M}{10} \tag{19.76′}$$

*for* $\sigma$ *outside a set of measure* $< e^{-M^{c_3}}$.

We first show how (†) is established inductively at scale

$$r < \exp(\log\log M_1)^3 \le r + 1 \tag{19.77}$$

According to (19.39), let

$$M_0 = \exp(\log\log M_1)^{10} \tag{19.78}$$

If we choose $r_0 \sim \log M_1$, clearly

$$\exp(\log\log M_0)^3 < r_0 < r$$

and (†) holds thus for $T^\sigma_{M_0,Q_0}$, $Q_0 = [-10M_0, 10M_0]^d$, for all $(\lambda, \lambda') \in \bigcup_{I_0 \in \Lambda_{r_0}} I_0$.

We use Lemmas 19.38 and 19.65, replacing $q$ by $q^{(r_0)}$ according to Remark (2) above and taking $(\lambda, \lambda') \in I_0$ for a fixed $I_0 \in \Lambda_{r_0}$.

Statements (19.76) and (19.76′) with $M = M_1$ then follow from (19.43) and (19.66), provided a subset $\mathcal{A} = \mathcal{A}(I_0)$ is removed from $I_0$, satisfying (19.75)

$$\text{mes}_{\sec}\mathcal{A}(I_0) < [\exp\exp(\log r_0)^2]^{-1} \tag{19.79}$$

and $\mathcal{A}$ composed of intervals of size

$$[\exp\exp(\log\log M_1)^3]^{-1} \sim e^{-r} \tag{19.80}$$

by (19.77).

Consider the curve $\Gamma_r$ defined by $\lambda' = \lambda + \varepsilon\varphi_r(\lambda)$ and

$$\bigcup_{I \in \Lambda_r} (\Gamma_r \cap I) \subset \bigcup_{I_0 \in \Lambda_{r_0}} (\Gamma_r \cap I_0)$$

It follows from (19.79) and (19.80) that there is $\Lambda'_r \subset \Lambda_r$ such that (†) holds for $M = M_1$ and $(\lambda, \lambda') \in \bigcup_{I \in \Lambda'_r} I$, with

$$\text{mes}_b\Big(\bigcup_{I \in \Lambda_r} (\Gamma_r \cap I) \setminus \bigcup_{I \in \Lambda'_r} (\Gamma_r \cap I)\Big) < A^{r_0^C}[\exp\exp(\log r_0)^2]^{-1}$$

$$< \frac{1}{M_1} \tag{19.81}$$

$$< [\exp\exp(\log r)^{1/3}]^{-1}$$

The next generation of intervals $\{I \in \Lambda_{r+1}\}$ of size $A^{-(r+1)^C}$ is then introduced as further refinement of $\{I \in \Lambda'_r\}$ to ensure, moreover, the required estimates for $T_N^{-1}$, $N = A^{r+1}$. To invert $T_N$, consider a paving of $[-N, N]^{b+d}$ by intervals $J$ of size $M = (\log N)^{C_6}$, for which we require

$$\|(R_J T R_J)^{-1}\| < e^{M^{1-}} \tag{19.82}$$

and

$$|(R_J T R_J)^{-1}(\xi, \xi')| < e^{-c|\xi - \xi'|} \text{ for } \xi, \xi' \in J, |\xi - \xi'| > \frac{M}{10} \tag{19.82′}$$

Thus $J = ([-M, M]^b + k) \times Q$, where $Q$ is a size-$M$ interval in $[-N, N]^d$. Since $T_{A^r}^{-1}$ already satisfies the required properties, it suffices to consider $\frac{N}{2A} = \frac{1}{2}A^r \le |k| \le N$.

We have thus

$$R_J T R_J = T_{M,Q}^{\sigma=k.\lambda'}$$

and we distinguish the cases $\min_{n\in Q} |n| > M$ and $Q = [-10M, 10M]^d = Q_0$. If $\min_{n\in Q} |n| > M$, invoke Lemma 19.13. Condition (19.16) becomes

$$\min_s |k.\lambda' - \sigma_s(\lambda, \lambda')| > e^{-M^{c_2}}$$

Define

$$r' = \exp(\log \log M)^3 < r \qquad (19.83)$$

Restricting $(\lambda, \lambda')$ to the graph $\Gamma_{r'}$, we need to exclude the set of $\lambda$'s for which

$$\min_Q \min_s |k.(\lambda + \varepsilon\varphi_{r'}(\lambda)) - \sigma_s(\lambda, \lambda + \varepsilon\varphi_{r'}(\lambda))| \le e^{-M^{c_2}} \qquad (19.84)$$

Since $\|\sigma_s\|_{\text{Lip}} \le CM$ and $|k| \gg M$, the measure of this set is at most

$$N^{b+d} e^{(\log M)^C} e^{-M^{c_2}} < e^{-\frac{1}{2}M^{c_2}} < A^{-r}$$

provided $c_2 C_6 > 1$.

For $Q = Q_0 = [-10M, 10M]^d$, we argue again as in Chapter 18, considering the semialgebraic sets

$$\mathfrak{S} = \{(\lambda, \lambda', \sigma) \in I' \times \mathbf{R} | T_{M,Q_0}^\sigma \text{ fails (19.76), (19.76')}\}$$

and

$$\mathfrak{S} \cap ((\Gamma_{r'} \cap I') \times \mathbf{R}) \qquad (19.85)$$

with $I' \in \Lambda_{r'}$. By (19.83) and (†), $T_{M,Q_0}^\sigma$ satisfies for all $(\lambda, \lambda') \in I'$ the bounds (19.76) and (19.76') for $\sigma$ outside a set of measure $e^{-M^{c_3}}$. The set (19.85) is semialgebraic of degree at most $(MA^{(r')^3})^C < \exp\exp(\log\log r)^4$. Take $C_6$ s.t. also $c_3 C_6 > 1$. Hence $e^{M^{c_3}} > e^{(\log N)^{1+}} > A^r$, and we may carry out the argument (18.29) and (18.35) (with $r_0$ replaced by $r'$). This provides a collection $\Lambda_{r+1}$ of size $A^{-(r+1)^C}$ intervals $I'$ in $\mathbf{R}^{2b}$, refining $\{I \in \Lambda_r'\}$, on which $T_N^{-1}, N = A^{r+1}$, satisfies the required bounds. By (19.81), we have, moreover,

$$\text{mes}_b\left(\bigcup_{I\in\Lambda_r} (\Gamma_r \cap I) \setminus \bigcup_{I'\in\Lambda_{r+1}} (\Gamma_r \cap I')\right) < [\exp\exp(\log r)^{1/3}]^{-1} \qquad (19.86)$$

Estimate (19.86) is weaker than (18.35) but equally suffices.

**Proof of Lemma 19.10.** Fix $B$. We need to show that if $n_1, \ldots, n_k \in \mathbf{Z}^d$ is a sequence of distinct elements s.t.

$$|n_j - n_{j+1}| + |n_j^2 - n_{j+1}^2| < B \text{ for } 1 \le j < k \qquad (19.87)$$

then $k < B^C$, for some $C = C(d)$. Denote

$$\Delta_j n = n_{j+1} - n_j$$

Assume $I \subset [1, \ldots, k]$ an interval s.t.

$$d_1 = \dim[\Delta_j n | j \in I'] = \dim[\Delta_j n | j \in I] \qquad (19.88)$$

for any subinterval $I' \subset I$ with
$$|I'| \geq K$$
Our aim is to obtain a lower bound on $K$ in terms of $|I|$ (and $B, d$). It follows from (19.87) that for $|j - j'| \leq K$,
$$|n_j - n_{j'}| + |n_j^2 - n_{j'}^2| < KB$$
$$2|(n_{j'} - n_j).n_j| \leq |n_j^2 - n_{j'}^2| + |n_{j'} - n_j|^2 < 2K^2B^2$$
Hence
$$|\Delta_{j'}n.n_j| < 2K^2B^2 \tag{19.89}$$
Fix $j \in I$, and let $j \in I' \subset I$, where $I'$ is an interval of length $K$. Thus (19.88) holds. Let $\xi_1, \ldots, \xi_{d_1}$ be linearly independent elements in $\{\Delta_{j'}n|j' \in I'\}$. Thus
$$[\xi_1, \ldots, \xi_{d_1}] = [\Delta_{j'}n|j' \in I'] = [\Delta_{j'}n|j' \in I]$$
Let $\xi = \sum_{s=1}^{d_1} c_s \xi_s \in [\xi_1, \ldots, \xi_{d_1}]$. Then
$$\sum_{s=1}^{d_1} c_s.\xi_s.\xi_{s'} = \xi.\xi_{s'} \quad (1 \leq s' \leq d_1)$$
and hence, by Cramer's rule,
$$|c_s| = \frac{|\det[(\xi_{s'}.\xi_{s''})_{1 \leq s' \leq d_1(s'' \neq s)}, (\xi.\xi_{s'})_{1 \leq s' \leq d_1}]|}{|\det[(\xi_{s'}.\xi_{s''})_{1 \leq s' \leq d_1, 1 \leq s'' \leq d_1}]|}$$
$$< d_1^{\frac{1}{2}}.|\xi|.B(d_1^{\frac{1}{2}}.B^2)^{d_1-1} = d_1^{\frac{d_1}{2}} B^{2d_1-1}|\xi| \tag{19.90}$$
since $|\xi_s| < B$ and $\det[\xi_{s'}.\xi_{s''}] \in \mathbf{Z}\backslash\{0\}$.

By (19.89),
$$|n_j.\xi_s| < 2K^2B^2 \text{ for } s = 1, \ldots, d_1$$
and (19.90) implies
$$|\xi.n_j| \leq \sum |c_s| |\xi_s.n_j| < 2d_1^{\frac{d_1}{2}+1} B^{2d_1+1} K^2|\xi| \tag{19.91}$$
Thus (19.91) holds for all $j \in I, \xi \in [\Delta_{j'}n|j' \in I]$. Therefore, if $j_1, j_2 \in I$, we get
$$|n_{j_1} - n_{j_2}| < 4d_1^{\frac{d_1}{2}+1} B^{2d_1+1} K^2 \tag{19.92}$$
Since $\{n_j\}$ consists of distinct elements in $\mathbf{Z}^d$, we may choose $j_1, j_2 \in I$ such that
$$|n_{j_1} - n_{j_2}| > |I|^{\frac{1}{d}}$$
Hence, from (19.92),
$$K > \frac{1}{2} d^{-\frac{d+2}{4}} B^{-d-\frac{1}{2}} |I|^{\frac{1}{2d}} \tag{19.93}$$
This means that there is an interval $I' \subset I$ for which
$$|I'| > d^{-(d+1)} B^{-d-1} |I|^{\frac{1}{2d}}$$
and
$$\dim[\Delta_j n|j \in I'] < d_1 = \dim[\Delta_j n|j \in I]$$
Starting from $I = [1, \ldots, k]$, iteration of the preceding at most $d$ times implies that
$$1 > (dB)^{-(d+1)\left(1+(2d)^{-1}+(2d)^{-2}+\cdots\right)} k^{\left(\frac{1}{2d}\right)^d} > (dB)^{-2(d+1)} k^{\left(\frac{1}{2d}\right)^d}$$
and hence
$$k < d^{(2d+2)^{d+1}} B^{(2d+2)^{d+1}}$$
This proves the lemma with $C_d = (2d + 2)^{d+2}$.

# Chapter Twenty

## Construction of Quasi-Periodic Solutions of Nonlinear Wave Equations

### 1. Formulation of the Problem

Consider NLW with periodic boundary condition ($x \in \mathbf{T}^d$) of the form

$$y_{tt} - \Delta y + \varepsilon F'(y) = 0 \tag{20.1}$$

where $F(y)$ is a polynomial in $y$.

The construction of time-periodic solutions was achieved in arbitrary dimension $d$ (see [B1]). Quasi-periodic solutions were so far only produced for $d = 1$ (see [B2], [Kuk], and [Wa] with 1D Dirichlet bc). In this chapter we will indicate how the methods described in the preceding chapter may be used to treat this problem in general dimension $d$. Rather than relying on amplitude-frequency modulation, extracting parameters from the nonlinearity, we discuss again nonlinear perturbations of a linear equation with parameters. Thus replace (20.1) by

$$y_{tt} + B^2 y + \varepsilon F'(y) = 0 \tag{20.2}$$

where $B$ is given by a Fourier multiplier defined as follows:

Let $0 \in \{n_1, \ldots, n_b\} \subset \mathbf{Z}^d$ be a set of distinguished modes. Let $B$ be defined by the multiplier $\{\mu_n\}_{n \in \mathbf{Z}^d}$, where

$$\begin{cases} \mu_{n_j} = \lambda_j > 0 & (1 \leq j \leq b) \\ \mu_n = |n| \text{ if } n \notin \{n_1, \ldots, n_b\} \end{cases} \tag{20.3}$$

and where again $\lambda = (\lambda_1, \ldots, \lambda_b)$ is a $b$-dim parameter.

Introducing as usual the speed $v = y_t$, (20.2) becomes

$$\begin{cases} y_t = v \\ v_t = -B^2 y - \varepsilon F'(y) \end{cases} \tag{20.4}$$

Denoting $z = B^{-1} v$ and $q = y - iz$, we thus obtain

$$\begin{cases} y_t = Bz \\ z_t = -By - \varepsilon B^{-1} F'(y) \end{cases} \tag{20.5}$$

or

$$\frac{1}{i} q_t = Bq + \varepsilon B^{-1} F'(\operatorname{Re} q) = B^{-1} \frac{\partial H}{\partial \bar{q}} \tag{20.6}$$

where

$$H(q) = \int \left[ \frac{1}{2} |Bq|^2 + 2\varepsilon F(\operatorname{Re} q) \right]$$

Denoting

$$H_1(q, \bar{q}) = 2F\left(\frac{q + \bar{q}}{2}\right)$$

the equation

$$\frac{1}{i}q_t = Bq + \varepsilon B^{-1}\frac{\partial H_1}{\partial \bar{q}} \tag{20.7}$$

will be treated following the method used in the previous chapter for NLS.
Write

$$q(x) = \sum_{n \in \mathbf{Z}^d} q_n e^{2\pi i n.x}$$

and

$$q_n(t) = \sum_{k \in \mathbf{Z}^b} \widehat{q_n}(k) e^{ik.\lambda' t}$$

Let

$$\mathcal{R} = \{(n_j, e_j) | j = 1, \dots, b\} \subset \mathbf{R}^{d+b}$$

be the set of resonant sites and specify again

$$\widehat{q_{n_j}}(e_j) = a_j \in \mathbf{R}_+ \tag{20.8}$$

(the case $\{a_j\}_{1 \le j \le b} \subset \mathbf{C}$ may be reduced to (20.8) by time shift).
Identifying $q$ and $\hat{q} = \hat{q}(n, k) = \hat{q}_n(k)$, rewrite (20.7) as

$$(k.\lambda' - \mu_n)\hat{q}(n, k) + \varepsilon\frac{1}{\mu_n}\frac{\widehat{\partial H_1}}{\partial \bar{q}}(n, k) = 0 \tag{20.9}$$

where $B_n = \mu_n \ne 0$, and make again a decomposition in $P$ and $Q$-equations.
Thus the $Q$-equations are

$$\lambda'_j - \lambda_j + \frac{\varepsilon}{a_j}\frac{1}{\lambda_j}\frac{\widehat{\partial H_1}}{\partial \bar{q}}(n_j, e_j) = 0 \qquad (1 \le j \le b) \tag{20.10}$$

(which permit us to express $\lambda' = \lambda'(\lambda)$), and the $P$ equations are now given by

$$\begin{cases} \mu_n(k.\lambda' - \mu_n)u(n, k) + \varepsilon\widehat{\frac{\partial H_1}{\partial \bar{q}}}(n, k) = 0 \\ \mu_{-n}(-k.\lambda' - \mu_{-n})v(n, k) + \varepsilon\widehat{\frac{\partial H_1}{\partial \bar{q}}}(n, k) = 0 \end{cases} \tag{20.11}$$

In the present case, the linearized operator $T$ is

$$T = T_q = D + \varepsilon S_q$$

where

$$D_{\pm,n,k} = \mu_{\pm n}(\pm k.\lambda' - \mu_{\pm n}) \tag{20.12}$$

We let $\lambda = \lambda_1, \dots, \lambda_b)$ vary in an interval $\Omega \subset \mathbf{R}^b$, and $\varepsilon$ is taken small enough.
In this setting, we have the analogue of Theorem 19.1 for (20.7).

The proof follows the same scheme explained in Chapters 18 and 19. There are some additional ingredients and issues that we will explain next.

## 2. Technical Details

Compared with the NLS case discussed in the previous chapter, the inversion of the $T_{M,Q}^{\sigma}$ operators requires certain modification of the arguments. In the present situation,

$$D_{\pm,n,k}^{\sigma} = \mu_n(\pm(k.\lambda' + \sigma) - \mu_n) \tag{20.13}$$

where $\mu_n = |n| \neq 0$ except for $n \in \{n_j | j = 1, \ldots, b\}$.

The most significant issue is to prove Lemma 19.13 in the present situation. This relies on the arithmetical Lemma 19.10, which is explicitly dependent on properties of the $\{\mu_n\}$. To obtain the required separation result, an additional nonlinear condition on $\lambda'$ will be imposed.

The next fact will substitute Lemma 19.10.

**Lemma 20.14.** *Let $B$ be a large number and assume $\lambda' \in \mathbf{R}^b$ satisfi es the condition*

$$|P(\lambda')| > B^{-C'}$$

*for all polynomials $P(X) \in \mathbf{Z}[X_1, \ldots, X_b], P(X) \neq 0$ of degree $\deg P < 10d$ and with coeffi cients $|a_\alpha| < B^C$. Consider a sequence $(\xi_j)_{1 \leq j \leq k}$ of distinct elements of $\mathbf{Z}^{b+d}$ s.t. for some $\sigma \in \mathbf{R}$, for all $j$*

$$|(\lambda'.k_j + \sigma)^2 - |n_j|^2| < 1 \text{ where } \xi_j = (k_j, n_j) \tag{20.15}$$

*and*

$$|\xi_j - \xi_{j-1}| < B \tag{20.16}$$

*Assume, moreover, that*

$$\max_n(\#\{1 \leq j \leq k | n_j = n\}) < B' \tag{20.17}$$

*Then*

$$k < (BB')^{C''} \tag{20.18}$$

(We will denote in what follows all constants possibly dependent on $d$ and $b$ by $C$.)

We prove Lemma 20.14 at the end of this chapter.

**Lemma 20.19.** *Lemma 19.13 holds provided $\lambda' \in \mathbf{R}^b$ satisfi es, moreover, the property*

$$|p(\lambda')| > M_0^{-C'} \tag{20.20}$$

*for any polynomial $p(X) \in \mathbf{Z}[X_1, \ldots, X_b], p(X) \neq 0$ of degree at most $10d$ and coeffi cients $\{a_\alpha\}$ bounded by $M_0^C$.*

**Sketch of the Proof.** We proceed again by multiscale analysis in $M_0$. Assume that the statement holds at scale $M_0$ for $T_{M_0,Q}^{\sigma}$, where $Q \subset \mathbf{Z}^d$ is a $M_0$-interval, $\min_{n \in Q} |n| > CM_0$.

Then take $M_1 = e^{M_0^{\frac{1}{2}c_2}}$ and let $Q \subset \mathbf{Z}^d$ be an $M_1$-interval s.t. $\min_{n \in Q} |n| > CM_1$. Call $(n, k)$ a good site if $T_{M_0,n+Q_0}^{\sigma+k.\lambda'}$, $Q_0 = [-M_0, M_0]^d$ satisfies (19.15) and (19.15'). Otherwise, $(n, k)$ is called a bad site.

Partition $[-M_1, M_1]^b \times Q$ in $(b+d)$-dim intervals $P$ of size $M_0$. Consider the collection $\mathcal{P}$ of intervals $P$ containing some bad site $(n, k)$. Thus

$$[-M_1, M_1]^b \times Q \equiv \Omega = \Omega_0 \cup \Omega_1$$

where

$$\Omega_1 = \bigcup_{P \in \mathcal{P}} P$$

and $(T_{\Omega_0}^\sigma)^{-1}$ is controlled by the resolvent identity.

Our next goal is to show that $\Omega_1$ has a separated cluster structure. Fix

$$B = M_1^\rho, \rho \text{ small enough}$$

Let $\{P_i | i = 1, \ldots, j\}$ be a sequence of distinct elements in $\mathcal{P}$ s.t.

$$\text{dist}(P_i, P_{i+1}) < B$$

Clearly, if $\xi$ is a bad site, there is some $\xi' = (n', k') \in [-M_1, M_1]^{b+d}$ for which

$$|\xi - \xi'| \lesssim M_0$$

and

$$\min_{\pm} |D_{\pm, n', k'}^\sigma| < \frac{1}{3}$$

Hence

$$\left| (k'.\lambda' + \sigma)^2 - |n'|^2 \right| < 1$$

We therefore may find a sequence $\xi_i = (n_i, k_i), 1 \le i \le j$, s.t.

$$\text{dist}(\xi_i, P_i) < CM_0$$

and hence

$$|\xi_{i+1} - \xi_i| < B + CM_0 < 2B$$

and

$$\left| (k_i.\lambda' + \sigma)^2 - |n_i|^2 \right| < 1$$

In order to apply Lemma 20.14, we need to check condition (20.17). By construction, it is clear that given $n \in [-M_1, M_1]^d$,

$$\#\{1 \le i \le j | n_i = n\}$$
$$< (CM_0)^d \max_{n'} [\#\{k \in [-M_1, M_1]^b | T_{M_0, n'+Q_0}^{\sigma + k.\lambda'} \text{ fails } (19.15), (19.15')\}]$$
$$(20.21)$$

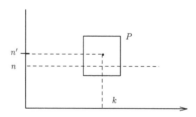

Since in particular from the assumption, $\lambda'$ satisfies the DC

$$\|k.\lambda'\| > M_1^{-C} \text{ for } k \in \mathbf{Z}^b, 0 < |k| < M_1$$

it follows again from the validity of Lemma 20.19 at scale $M_0$ that

$$\#\{\ldots\} < e^{(\log M_0)^C}$$

Hence

$$(20.21) < (CM_0)^d \, e^{(\log M_0)^C}$$

and we may take

$$B' = e^{(\log M_0)^C}$$

in (20.17). Applying Lemma 20.14, we conclude that

$$j < \left(B.e^{(\log M_0)^C}\right)^C < M_1^{\rho C+} < M_1^{\frac{1}{100}}$$

for appropriate choice of $\rho$.

This proves that

$$\Omega_1 = \bigcup_\beta \Omega_{1,\beta}$$

where

$$\operatorname{diam} \Omega_{1,\beta} < M_1^{\frac{1}{100}}$$

and

$$\operatorname{dist}(\Omega_{1,\beta}, \Omega_{1,\beta'}) > M_1^\rho \text{ for } \beta \neq \beta'$$

We continue the argument as in Lemma 19.13. Let thus $\widetilde{\Omega}_{1,\beta}$ be an $M_1^{\frac{3}{4}\rho}$-neighborhood of $\Omega_{1,\beta}$. Assume $\sigma > 0$. Since $\min_{n \in Q} |n| > CM_1$, clearly,

$$|D_{-,n,k}^\sigma| = |n| \, |k.\lambda' + \sigma + |n|| > M_1(CM_1 - 0(M_1)) > M_1^2$$

and

$$\|(R_- T_{\widetilde{\Omega}_{1,\beta}}^\sigma R_-)^{-1}\| < M_1^{-2}$$

The $\{\sigma_i\}$-functions are introduced as in Lemma 19.13. Thus, if

$$\min_i |\sigma - \sigma_i(\lambda, \lambda')| > e^{-M_1^{\rho/2}}$$

we get

$$\|(T_{\widetilde{\Omega}_{1,\beta}}^\sigma)^{-1}\| < e^{M_1^{\rho/2}} \text{ for all } \beta$$

The resolvent identity then gives the desired bound on $(T_\Omega^\sigma)^{-1}$.

The proof of the analogue of Lemma 19.38 is essentially the same. The only difference is that (19.49) gets replaced by the (weaker) restriction

$$|k - \hat{k}| + | \, |n| - R| < M^{1+}$$

and (19.56) becomes

$$\#\Lambda < M^C . R^{d-1}$$

The argument based on the dichotomy $R < M_1^{\frac{1}{100d}}, R > M_1^{\frac{1}{100d}}$ remains identical. The remainder of the argument is the same.

**3. Proof of Lemma 20.14.** Since, by density, we may take $\sigma = k.\lambda'$ for some $k \in \mathbf{Z}^b$ and the problem is invariant under translation, we may assume $\sigma = 0$. Thus the hypothesis (20.15) becomes

$$|(\lambda'.k_j)^2 - |n_j|^2| < 1 \quad (1 \le j \le k) \tag{20.22}$$

Define the operators

$$T_{\pm} : \mathbf{R}^{b+d} \to \mathbf{R}^{d+1} : (k, n) \mapsto (n, \pm k.\lambda')$$

Thus, for $\xi = (k, n)$,

$$|n|^2 - (\lambda'.k)^2 = T_+\xi.T_-\xi$$

The hypothesis on $\lambda'$ is exploited as follows.

**Lemma 20.23.** Let $(\xi_s)_{1 \le s \le d_1} (d_1 \le d), \xi_s = (k_s, n_s) \in \mathbf{Z}^{b+d}, |\xi_s| < B$, s.t.

$$\dim[\xi_s | 1 \le s \le d_1] = d_1$$

and

$$\dim[n_s | 1 \le s \le d_1] \ge d_1 - 1$$

Then, under the assumption on $\lambda'$, we have

$$|\det(T_+\xi_s.T_+\xi_{s'})_{1 \le s, s' \le d_1}| > B^{-C} \tag{20.24}$$

and

$$|\det(T_+\xi_s.T_-\xi_{s'})_{1 \le s, s' \le d_1}| > B^{-C} \tag{20.24'}$$

**Proof of Lemma 20.22.** We prove (20.24′). Define

$$P(X) = \det\left(n_s.n_{s'} - (k_s.X)(k_{s'}.X)\right) \in \mathbf{Z}[X_1, \ldots, X_b]$$

which is a polynomial of degree $\le 2d_1$ and coefficients

$$|a_\alpha| \le d_1!(1 + b^2)^{d_1}\left(\sum |\xi_s|^2\right)^{d_1} < CB^{2d_1} \tag{20.25}$$

We claim that $P(X) \ne 0$. If $P(X) = 0$, then also

$$0 = P(iX) = \det\left(n_s.n_{s'} + (k_s.X)(k_{s'}.X)\right)$$

implying that the vectors $(n_s, k_s.X) \in \mathbf{R}^{d+1}$ $(1 \le s \le d_1)$ are linearly dependent for all $X \in \mathbf{R}^b$.

Since from assumption

$$\dim[n_s | 1 \le s \le d_1] = d_1 - 1$$

there is a unique (up to multiples) vector $(c_s)_{1 \le s \le d_1} \in \mathbf{R}^{d_1} \setminus \{0\}$ s.t.

$$\sum c_s n_s = 0 \quad \text{in } \mathbf{R}^d$$

and therefore also satisfying

$$\left(\sum c_s k_s\right).X = 0 \text{ hence } \sum c_s k_s = 0 \text{ in } \mathbf{R}^b$$

This contradicts the linear independence of $(\xi_s)_{1 \le s \le d_1}$. Thus $P(X) \ne 0$ and (20.24′) follow from (20.25) and the assumption on $\lambda'$.

We now return to the proof of Lemma 20.14. Denote

$$\Delta_j \xi = \xi_j - \xi_{j-1}$$

Fix $K \in \mathbf{Z}_+$ and let $|j - j'| \le K$. From (20.16) and (20.22)

$$\begin{aligned} 2|T_+\xi_j.T_-(\xi_{j'} - \xi_j)| \quad &\le |T_+\xi_j.T_-\xi_j| + |T_+\xi_{j'}.T_-\xi_{j'}| \\ &+ |T_+(\xi_{j'} - \xi_j).T_-(\xi_{j'} - \xi_j)| \\ &\le 2 + C\|\xi_{j'} - \xi_j\|^2 < CK^2 B^2 \end{aligned}$$

Hence, by subtraction, also

$$|T_+\xi_j.T_-\Delta_{j'}\xi| < CK^2 B^2 \tag{20.26}$$

Let $I \subset [1, k]$ be an interval and $K \in \mathbf{Z}_+$ s.t.

$$\begin{aligned} d_1 \quad &\equiv \dim[T_+\Delta_j\xi | j \in I] \\ &= \dim[T_+\Delta_j\xi | j \in I'] \end{aligned}$$

for all $I' \subset I, |I'| \ge K$. Then take $j \in I$, and let $I', |I'| = K$ be an interval s.t. $j \in I' \subset I$. Denote

$$E_{\pm} = [T_{\pm}\Delta_{j'}\xi | j' \in I'] = [T_{\pm}\Delta_{j'}\xi | j' \in I]$$

and $\{e_1, \ldots, e_{d_1}\} \subset \{T_+\Delta_{j'}\xi | j' \in I'\}$ a basis for $E_+$.

By (20.26) and the choice of $I'$,

$$|T_-\xi_j.e_s| < CK^2 B^2 \quad (1 \le s \le d_1) \tag{20.27}$$

Since the vectors $\{e_s = (\Delta_{j_s} n, \Delta_{j_s} k.\lambda')\}_{1 \le s \le d_1}$ are linearly independent in $\mathbf{R}^{d+1}$, it follows that $\{\Delta_{j_s}\xi | 1 \le s \le d_1\}$ are linearly independent and $\dim[\Delta_{j_s} n | 1 \le s \le d_1] \ge d_1 - 1$.

Since $\|\Delta_{j_s}\xi\| < B$, (20.24) implies that

$$\begin{aligned} |\det(e_s.e_{s'})| \quad &= |\det(T_+\Delta_{j_s}\xi.T_+\Delta_{j_{s'}}\xi)| \\ &> B^{-C} \end{aligned} \tag{20.28}$$

From (20.28) and Cramer's rule,

$$B^C \|\sum_{s=1}^{d_1} c_s e_s\| \ge \max |c_s| \text{ for all } \{c_s\}$$

and hence, by (20.27),

$$|T_-\xi_j.v| < B^C K^2 \tag{20.29}$$

for any $v \in E_+, \|v\| \le 1$. Equivalently,

$$|T_+\xi_j.w| < B^C K^2 \tag{20.30}$$

for any $w \in E_-, \|w\| \leq 1$. Recall that $j \in I$ was an arbitrarily chosen element. Let $\zeta = \xi_j - \xi_{j'}$ for $j, j' \in I$.

It follows from (20.30) that

$$|T_+\zeta.w| < B^C K^2 \tag{20.31}$$

for any $w \in E_-, \|w\| \leq 1$.

Since $T_+\zeta \in E_+$, we may write

$$T_+\zeta = \sum_{s=1}^{d_1} c_s e_s$$

$$\left| \sum c_s e_s.\bar{e}_{s'} \right| = |T_+\zeta.\bar{e}_{s'}| < B^C K^2 \tag{20.32}$$

where $\bar{e}_s = T_-\Delta_{j_s}\xi$.

By (20.24'), also

$$|\det(e_s.\bar{e}_{s'})| = |\det(T_+\Delta_{j_s}\xi.T_-\Delta_{j_s}\xi)|$$
$$> B^{-C} \tag{20.33}$$

From (20.32) and (20.33) and Cramer,

$$|c_s| < B^C K^2$$

Hence

$$\|n_j - n_{j'}\| \leq \|T_+(\xi_j - \xi_{j'})\| = \|T_+\zeta\| < B^C K^2 \tag{20.34}$$

It follows that

$$\text{diam}\, \{n_j | j \in I\} < B^C K^2$$

Hence

$$\#\{n_j | j \in I\} < B^C K^{2d} \tag{20.35}$$

Since we also assume (20.17), (20.35) implies that

$$|I| < B' B^C K^{2d} \tag{20.36}$$

This means that there is an interval $I' \subset I$ satisfying

$$|I'| > (B')^{-\frac{1}{2d}} B^{-C} |I|^{\frac{1}{2d}} \tag{20.37}$$

for which

$$\text{dim}[T_+\Delta_j\xi | j \in I'] < \text{dim}[T_+\Delta_j\xi | j \in I]$$

Starting from $I = [1, k]$, at most $d+1$ iterations of the preceding clearly lead to the following conclusion:

$$1 > (B')^{-\frac{1}{d}} B^{-C} k^{(\frac{1}{2d})^d} \tag{20.38}$$

This proves (20.18) and Lemma 20.14.

# References

[1] J. Bourgain. Construction of periodic solutions of nonlinear wave equations in higher dimension, *GAFA* 5 (1995), 629–639.

[2] J. Bourgain. Nonlinear Schrödinger equations. In *Hyperbolic equations and frequency interactions*, IAS/Park City Math. Ser. 5, Amer. Math. Soc., Providence, RI, 1999, pp. 3–157.

[3] S. Kuksin. Nearly integrable infinite-dimensional Hamiltonian systems, *Lecture Notes in Math.*, Springer-Verlag, Berlin, 1993, p. 1556.

[4] C. Wayne. Periodic and quasi-periodic solutions of nonlinear wave equations via KAM theory, *CMP* 127 (1990), 479–528.

# *Appendix*

**Strongly Mixing Potentials**

We first recall the Figotin-Pastur [F-P] formalism, based on the representation of Schrödinger matrices in polar coordinates.

Consider the discrete Schrödinger operator

$$(H_\lambda \psi)_n = \psi_{n+1} + \psi_{n-1} + \lambda v_n \psi_n \tag{1}$$

where $v_n$ is a sequence of real numbers and $\lambda > 0$. We consider (1) both on the integer lattice $\mathbf{Z}$ or on the half-line $\mathbf{Z}^+ \cup \{0\}$. At this stage, we do not specify $\{v_n\}$. Fix some $\delta > 0$, and restrict the energy to

$$\delta < |E| < 2 - \delta \tag{2}$$

Then define $\kappa \in (0, \pi)$ and $V_n$ by

$$E = 2 \cos \kappa \tag{3}$$

$$V_n = -\frac{v_n}{\sin \kappa} \tag{4}$$

Next, let $\psi$ be a solution of $H\psi = E\psi$ on the half-line $\mathbf{Z}^+ \cup \{0\}$. Thus

$$\begin{pmatrix} \psi_{n+1} \\ \psi_n \end{pmatrix} = \begin{pmatrix} E - \lambda v_n & -1 \\ 1 & 0 \end{pmatrix} \begin{pmatrix} \psi_n \\ \psi_{n-1} \end{pmatrix} \quad \text{for any } n = 1, 2, \ldots. \tag{5}$$

The coordinate change

$$Y_n = (\psi_n - \cos \kappa \psi_{n-1}, \sin \kappa \psi_{n-1}) \tag{6}$$

transforms (5) into

$$Y_{n+1} = \begin{pmatrix} \cos \kappa & -\sin \kappa \\ \sin \kappa & \cos \kappa \end{pmatrix} Y_n + \lambda V_n \begin{pmatrix} \sin \kappa & \cos \kappa \\ 0 & 0 \end{pmatrix} Y_n \tag{7}$$

Introducing polar coordinates,

$$Y_n = \rho_n (\cos \varphi_n, \sin \varphi_n) \tag{8}$$

Equation (7) implies

$$\rho_{n+1} \begin{pmatrix} \cos \varphi_{n+1} \\ \sin \varphi_{n+1} \end{pmatrix} = \rho_n \begin{pmatrix} \cos(\varphi_n + \kappa) + \lambda V_n \sin(\varphi_n + \kappa) \\ \sin(\varphi_n + \kappa) \end{pmatrix} \tag{9}$$

$$\cot g \varphi_{n+1} = \cot g (\varphi_n + \kappa) + \lambda V_n \tag{10}$$

$$\rho_{n+1}^2 = \rho_n^2 \left(1 + \lambda V_n \sin \left(2(\varphi_n + \kappa)\right) + \lambda^2 V_n^2 \sin^2 (\varphi_n + \kappa)\right) \tag{11}$$

From (11) with $\rho_1 = 1$, we get

$$\frac{1}{N} \log \rho_N(\theta) = \frac{1}{2N} \sum_1^N \log\big(1 + \lambda V_n \sin 2(\varphi_n + \kappa) + \lambda^2 V_n^2 \sin^2(\varphi_n + \kappa)\big) \tag{12}$$

$$= \frac{\lambda^2}{8N} \sum_1^N V_n^2 \tag{13}$$

$$+ \frac{\lambda}{2N} \sum_1^N V_n \sin 2(\varphi_n + \kappa) \tag{14}$$

$$- \frac{\lambda^2}{4N} \sum_1^N V_n^2 \cos 2(\varphi_n + \kappa) \tag{15}$$

$$+ \frac{\lambda^2}{8N} \sum_1^N V_n^2 \cos 4(\varphi_n + \kappa) + O(\lambda^3) \tag{16}$$

Letting

$$\begin{aligned} d\zeta_n &= e^{2i\varphi_n} \\ \mu &= e^{2i\kappa} \end{aligned} \tag{17}$$

one verifies that (10) is equivalent to

$$\zeta_{n+1} = \mu\zeta_n + \frac{i\lambda}{2} V_n \frac{(\mu\zeta_n - 1)^2}{1 - \frac{i\lambda}{2} V_n(\mu\zeta_n - 1)} \tag{18}$$

The idea is that (13) produces the main term and (14) to (16) appear as error terms for $N \to \infty$ (assuming the potential $\{v_n\}$ sufficiently mixing). Notice that (18) permits us to recover the phases $\{\varphi_n\}$ recursively, but tracking their distribution from this formula is not obvious.

Figotin and Pastur [F-P], Theorem 14.6, used this formalism to show that for small $\lambda$ the Lyapounov exponent $L(\lambda, E)$ obeys the expansion

$$L(\lambda, E) = \frac{\lambda^2 \mathbf{E}(v_0^2)}{2(4 - E^2)} + O(\lambda^3) \tag{19}$$

provided the potentials are identically distributed independent random variables with zero mean (notice that a change of variables, as the one given by (6), does not change the limit $\lim_{N \to \infty} \frac{1}{N} \log(|\psi_{n-1}| + |\psi_n|) = L(E)$.) The constant in $O(\lambda^3)$ depends on $E$ but remains bounded on intervals of the form $[-2 + \delta, -\delta] \cup [\delta, 2 - \delta]$. Observe that $\zeta_{n+1}$, hence $\varphi_{n+1}$, only depends on $V_1, \ldots, V_n$ (by (18)). In particular, if $\{v_n\}$ are i.i.d., $V_n$ and $\varphi_n$ are independent random variables. Taking expectations in (13) to (16), one thus obtains (19).

Suppose next that $v_n = v_n(x) = F(T^n x)$, where $T : X \to X$ is an ergodic transformation on some probability space $(X, \mu)$. If $T$ is assumed strongly mixing, Chulaevsky and Spencer showed in [C-S] that (19) remains valid–with $\mathbf{E}(v_0^2)$ replaced by the expression

$$\sigma(E) = \sum_{\ell=-\infty}^{\infty} e^{2i\kappa\ell} \langle F, F(T^\ell \cdot) \rangle \tag{20}$$

This is accomplished by iteration of (18) and exploiting the decay of the correlations. Typical examples are

(i) **The Period Doubling Map.**

Thus

$$v_n = \lambda F(2^n x) \quad (x \in \mathbf{T}) \tag{21}$$

where $F$ is $2\pi$-periodic and $\int F = 0$.

(ii) **Hyperbolic Toral Automorphisms** $A : \mathbf{T}^2 \to \mathbf{T}^2$.

In this case,

$$v_n = \lambda F(A^n x), x \in \mathbf{T}^2 \tag{22}$$

and again $F$ is $2\pi$-periodic, and $\int F = 0$.

In both examples (i) and (ii), we assume $F$ a $C^1$-function (this suffices for our purpose). In these cases it was shown in [C-S] that $L(\lambda, E)$ admits an asymptotic expansion

$$L(\lambda, E) = \lambda^2 c_0(E) + o(\lambda^3) \tag{23}$$

for small $\lambda$ and $0 < |E| < 2$. Here $c_0(E)$ is some function of $E$ that depends on $F$. In [B-G] we basically develop further the [C-S] analysis. Returning to (14) to (16), it turns out that (14) is the most difficult term to control (because it is linear in $\lambda$). Now in examples (21) and (22), it is possible with some work to control this sum by basically martingale difference sequences. Using the standard deviation estimates in martingale theory, this allows us to obtain an LDT

$$\mathrm{mes}\,[x| \,|\frac{1}{N} \log \|M_N(x, E)\| - L_N(E)| > \lambda^{5/2}] < \exp(-C_\lambda n) \tag{24}$$

Combining (24) with some of the methods developed in the quasi-periodic case, in particular in Chapters 6, 7, and 10, the following further results are obtained in [B-S].

**Theorem.** *Let* $H_\lambda(x) = \lambda F(T^n x)\delta_{nn'} + \Delta$ *with $T$ and $F$ as in examples (21) and (22). Fix the energy range $\delta < |E| < 2 - \delta$, and let $\lambda > 0$ be small enough. Then the IDS of $H_\lambda$ is Hölder continuous. Moreover, in example (21), $H_\lambda(\theta)$ satisfi es Anderson localization (i.e., pure point spectrum with exponentially localized states) for almost all $\theta$. In example (22), AL holds on any interval $I_0 \subset [-2 + \delta, -\delta] \cup [\delta, 2 - \delta]$ on which $\sigma(E) > 0$, where $\sigma(E)$ is given by (20).*

See [B-S] for further details.

**Remarks.**

**1.** The Figotin-Pastur method uses the fact that $\lambda$ is small. In analogy with the random models, one should expect the theorem to hold for arbitrary $\lambda \neq 0$ (considering, for instance, the model (21)). It does not seem obvious, however, to adjust the Furstenberg-Lepage technique based on Perron-Frobenius to a quasi-random setting.

**2.** In the same spirit, one may ask whether the IDS for the operator $H_\lambda$ considered above is smooth.

**3.** One could hope that the Figotin-Pastur method also may be applicable in the context of weakly mixing transformations, for instance, given by a skew shift. Thus take

$$H_\lambda(x) = \lambda \cos(x_1 + nx_2 + \frac{n(n-1)}{2}\omega)\delta_{nn'} + \Delta \tag{25}$$

Can one use the Figotin-Pastur approach to prove positivity of the Lyapounov exponent of (25) for $\lambda > 0$?

Some results on Lyapounov exponent and localization of (25) for small $\lambda$ may be found in [B1] and [B2].

# References

[B1]  J. Bourgain. Positive Lyapounov exponents for most energies, *GAFA* 1745 (2000), 37–66.

[B2]  J. Bourgain. On the spectrum of lattice Schrödinger operators with deterministic potential, *J. Analyse* 87 (2002), 37–75 and 88 (2002), 221–254.

[B-S]  J. Bourgain, W. Schlag. Anderson localization for Schrödinger operators on **Z** with strongly mixing potentials, *CMP* 215 (2000), 143–175.

[C-S]  V. Chulaevsky, T. Spencer. Positive Lyapounov exponents for a class of deterministic potentials, *CMP* 168 (1995), 455–466.

[F-P]  A. Figotin, L. Pastur. Spectra of random and almost periodic operators, *Grundlehren der mathematischen Wissenshaften*, Springer, Berlin 1992, p. 297.

Milton Keynes UK
Ingram Content Group UK Ltd.
UKHW020619300723
426013UK00005B/88